THE
SCIENCE BEHIND
CLIMATE CHANGE
AND
NATURAL LAW
PRINCIPLE
OF
INVINCIBILITY

TABLE OF CONTENTS

1
INTRODUCTION

The issue of climate change has been argued for decades and has become the most controversial global concern of our time. Climate deniers say global warming is a cyclical pattern of climates changing in the long history of the planet and some say the earth's temperature is cooling even with evidence of global warming. Many agree climate change and global warming is man-made. Several studies have been made by scientists from different fields and some state varied contrasting explanations. They have describe climate change, yet there are still many questions referring to the unpredictable and dramatic temperature changes that characterize behaviors of air particles that cause destructive typhoons, floods and droughts constantly prevailing on earth seasons after season.

The impact of climate change is contained in the physics law of thermodynamics referring to the behavior of air, when warmed cause droughts and fires, when cooled, brings floods[1] causing magnitudes of devastations on the planet which create tremendous problems in the lives of many people. For these reasons, the cause, prevention and control of climate change are closely studied in the hope of finding a solution to prevent catastrophes that constantly beset man's surrounding environment.

An unconventional and unique scientific observation shows that climate change can be controlled. These observations have been described, proven and tested more than fifty times around the world. It clarified the cause and remedy for climate change and control of floods, droughts and forest fires. The concept described herein has its origin from the ancient Vedic civilization of India brought to light by the late Maharishi Mahesh Yogi, who was a physicist and at the same time he was a highly revered Indian master. He graduated with a physics degree from Allahabad University in India in the early 1940. Maharishi was the founder of Transcendental Meditation, a popular alternative system of well-being and health care that has a profound and significant effect in the environment when established in large groups together. It describes the behavior of the finest elementary particles of nature which create invincible fields that shield the earth, like the earth's magnetic field, it shields the earth from solar flares,

[1] *See Chapter 7, Science Behind Climate Change*

otherwise, no life form could exist on earth. Compelling proofs already exists illustrating climate change can be controlled and balanced at the level of the Unified Field[2] occurring within the most silent level of the human mind reached during the practice of Maharishi's Transcendental Meditation and Yogic flying in large groups together. It is like one photon particle or one magneto particle, they have no significant strength standing alone. In contrast, large groups of photon particles create a larger, more powerful photon fields or more magnets joined together create greater magnetic field. The discovery of the Unified Field has ushered in the most significant knowledge referring to the transformation of man and his environment brought to light in this generation. Large groups of human brain placed at the level of silence create the invincible Unified Field. Invincibility means the ability of the human mind to shield his environment from negative trends. No negativity can penetrate the field created by the finest particles found in nature, also found within the human brain at the level of silence in large groups together.

In the early nineteenth century, Albert Einstein, a prominent scientist from the west, extensively pursued the Unified Field to the very end of his life but failed to locate its existence. Maharishi was able to validate the existence of the Unified Field because of Maharishi's background in the eastern wisdom of meditation that he learned from his master, Guru Dev, from the Himalayas. With a physics degree, Maharishi correlated the "sap" of the universe referring to the unseen, non-material properties of the finest elementary particles of nature[3] that described their invincible character enlivened at the most silent state of the human mind in large groups together. The application of the Unified Field found to occur within the human mind is an effective and instant solution to climate change, in large groups together at the level of silence. Like pouring a pot of cold water instantly cools off another pot of boiling water.

Einstein was the fore runner in the breaking down of the atom like the hydrogen and oxygen. Einstein postulated that finer atomic particles become powerful in group form, like the containment of hydrogen particles in mega tons can create a super bomb that can cause a destructive huge blast, in contrast, a single hydrogen particle on its own, hold no great power, unlike that which can be displayed in their collective form. As atoms break down to finer particles, from

---

[2] *The Discovery of the Unified Field, see Chapter 8*

[3] *The Science behind Climate Change, see Chapter 7*

hydrogen, oxygen to neutrons, protons, neutrinos, photinos, quarks, fermions and to finer particles, they create a field, like the neutrons create the electric field, magneto particles create the magnetic field, the graviton particles create the gravitational field hovering over each of the planets. Elementary particles of nature display the character of invincibility available in nature, the finer particles become, the field they create becomes invincible. Since the Unified Field contains the finest particle of nature, it creates the most powerful field, the Unified Field that is invincible from any negative trends occurring on earth including climate change. Like the magnetic field of the earth, it creates a field that shields the earth from solar flares. The human mind has the capability to create the Unified Field that shields the earth from any negativity occurring on earth. With the Unified Field established in large groups together, climates balance, terrorist are contained, conflicts and war end. Peace and prosperity prevail on earth. Maharishi said that the basis of life is like the root of the tree which remains unseen from the ground. When the leaves are decaying, water the root and new leaves will grow and replace the dry leaves. Maharishi has said that the Unified Field is the unseen basis of all life and the root of all creation, also enlivened within the silent level of the human mind. The solutions to problems of man can be found at the root. The solution to climate change lies within the invincible character of the Unified Field found enlivened within the human mind at the level of silence. Maharishi defined natural law principle of invincibility[4] applicable to climate change, control of floods, droughts and other natural disasters. The Maharishi Effect,[5] so named in honor of the late Maharishi Mahesh Yogi refers to natural law principle of invincibility.

I had the opportunity to have worked briefly with Maharishi when he was alive. I included herein events illustrating the effects of the Unified Field in the transformation of climates. Scientific evidences already exist and too important to miss. Maharishi's natural law principle of invincibility shield man from problems and suffering. This knowledge needs to be known and applied to control global warming, prevent floods, tornadoes and hurricanes caused by extreme and drastic changes of the weather constantly occurring in our environment.

Erlinda Case, Author

---

[4] *Natural Law Principle of Invincibility, see Chapter 14*

[5] *The Maharishi Effect, see Chapter 4*

2

THE CONCEPT OF CLIMATE CHANGE

Scientists describe climate change as the result of fossil fuel emissions that form the CO_2 layer hovering above the surface of the earth. The CO_2 cannot disperse into space because it cannot penetrate the invincible electromagnetic and gravitational field that envelops the earth. It stays below the earth's invincible field[6] and traps the heat reflected from the earth. This layer has been increasing and warms the air in the atmosphere causing fires and droughts or when cooled at the onslaught of winter, it cools the warm air from the sea that transforms to liquid at lower temperature causing torrential rains causing massive floods. The law of thermodynamics describes the natural behavior of air, when air is intensely warmed, it evaporates, when cooled, it transforms to water, further cooled turns to solid ice.

CO_2 has been identified as the major cause of climate change and referred to as the greenhouse effect. The thicker and larger the CO_2 layer becomes, the larger drought it can produce during warm season or when the cold season starts to set in, it can produce massive flood waters on earth. When sunlight strikes the earth, the earth absorbs heat and reflects it back to the atmosphere. The CO_2 beneath the layer of clouds above the earth catch and traps the heat. The greater concentration of CO_2 that is trapped in the atmosphere, the thicker and denser the greenhouse layer becomes, naturally the greater it can absorb heat and produce warmer temperature in the atmosphere, evaporates the air that protects the earth causing drought, or vice versa, it can create more water when cooled. CO_2 is heavier than air and so the CO_2 layer stays lower below the level of the clouds from the surface of the earth described as the earth's ozone layer. When warmed, it evaporates and leaves an open space for more heat coming from the sun. Satellite photo from the International Space Station shows the hole above the north pole created by the depletion of the ozone layer.

With the onslaught of the cold season, it absorbs more cold air due to the cooling of the atmosphere and sucks the air from the warm ocean waters causing tremendous air turbulence with great speed resulting in imbalance of the air equilibrium in the earth's

---

[6] *See Chapter 14, Natural Law Principle of Invincibility*

atmosphere causing strong hurricanes, typhoons and when air is cooled, it causes massive rain waters to flood the earth.

The behavior and character of the elementary particles of nature describes natural law principle of invincibility, like the magneto particles are finer than the hydrogen, oxygen or the carbon particle. The graviton particle is finer than the magneto particles. Fine particles of nature create a field, like the graviton particles create the gravitational field hovering around the earth's atmosphere, the magneto particles create the earth's magnetic field, or the neutron particles create the earth's electric field. Scientists found that the electric field is the same as the magnetic field, like connect a magnet to a coil, magnetic field is produced or vice versa, when a coil is connected to a magnet, electricity is produced, it becomes electro-magnetic field instead. The finer particles of nature become, they create stronger fields of invincibility. Invincibility means the inability of dissimilar or impure particles of greater size to penetrate pure fields, like the electro- magnetic or pure graviton field.

Maharishi's discovery of natural law principle of invincibility brought to light the Unified Field. The finest particles of nature create the Unified Field. Since the Unified Field contains the finest of all particles of nature, it is invincible from any negative trends occurring on earth. The CO_2 layer cannot penetrate the earth's magnetic field because the magneto particles are finer than the CO_2 particles, therefore the magnetic field is invincible from the CO_2 particles, so instead the CO_2 accumulates, become dense and get trapped beneath the magnetic field, at the greenhouse layer unable to disperse into space. They absorb heat reflected from the earth which increases gradually and depletes the ozone layer, warms the atmosphere and eventually cause global warming.

Before the industrial revolution that started about the year 1750, carbon dioxide was measured relatively low in parts per million, then rose to current data of nearly 400 parts per million of volume of concentration in the earth's atmosphere indicating that indeed global warming is occurring and is caused by the increase of human activity in the manufacture and use of fossil fuels that releases large amounts of CO_2 in the atmosphere. Scientist claim that the amount of CO_2 concentration has now grown more dense and has trapped larger amount of heat shown by severe droughts and devastating floods that has increasingly grown in number, sometimes three typhoons circulating at the same time and at the same place with greater intensity much more happening today around the world than that occurred prior to the beginning of the industrial revolution.

Greenhouse gas emissions come from man-made rapid commercial and manufacturing activities from the world's major industrialized nations in the processing of coal, natural gas and petroleum including various fossil fuels, cement and other related industries. Emissions from traffic and moving vehicles have increased through the years because the transport needs of the world's increasing population that has largely grown contributing to CO_2 concentration from vehicle exhaust emissions.

Plants and animals including man contribute to the emission of CO_2 thru natural aerobic respirations, but these emissions are balanced by the natural process of ecosystem where CO_2 is absorbed by large rain forests and oceans under normal low temperature fluctuations. The rapid increase of man-made greenhouse gases has imbalanced the earth's natural ecosystem.

CO_2 has no liquid state under normal pressure, in contrast to air which has the three states, solid, liquid and gas. Air, H_20 turns solid ice at zero degrees centigrade and carbon dioxide turn solid "dry ice" at -78 degrees centigrade. This is the reason CO_2 stays much longer in the atmosphere, as long as 35 to 95 years and sometimes extend thousands of years. Air can stay just a few years in the atmosphere because it can turn liquid and fall to earth as rain at lower temperatures, like in the evening, when the sun is farther away and out on the other side of the earth, the atmosphere cools off, the air turns to cloud and then turn to water and falls to earth as rain or as morning dews down on earth. CO_2 is soluble in water and it can mix with water falling down as rain.

The earth tilts away from the sun during winter and cools the earth's atmosphere. During winter, the CO_2 layer simultaneously cools too and ice caps are created at the north pole that increase more cold air on earth. During summer, the earth tilts towards the sun and warms the earth and if the greenhouse layer has increased, it correspondingly warms up the earth's atmosphere more. Global warming is occurring because greenhouse gases, mostly made of CO_2 has gradually accumulated in the atmosphere thru the years that absorbs heat emitted from the earth. The CO_2 cannot break through the air above it and thru the earth's magnetic field because it is heavier than air and the magneto particles are much finer in form and create the magnetic field of the earth that is impenetrable from solar flares or excess heat from the sun. The CO_2 cannot penetrate the earth's magnetic field. Instead, the CO_2 layer warms up the earth's atmosphere and then melts the ice caps at the north pole consequently increasing the water levels of the sea.

When the cool season sets in, it cools the CO_2 layer, then it sucks the warm air evaporating from the ocean waters causing turbulence in the air. Cold air turns to water, then storms occur and is swept on land causing floods, destroying bridges, breaking levies resulting in massive flooding and devastations on earth.

Climates are controlled by temperature conditions season after season. Winter season brings cold air. Summer season brings warm temperature. When the layer of the CO_2 that has absorbed the heat from the surface of the earth heats up the clouds and evaporates the ozone layer, droughts on land occur. Rain occurs when air is cooled whenever air turns to clouds as the temperature goes down and then rain falls as water on earth. Air turns to liquid water at lower temperatures and falls as rain on earth. Since CO_2 is soluble in water, rain mixes with the CO_2 and falls back to earth. Less rain occurs when the layer of CO_2 particles increases, it heats up the clouds covering the earth, evaporates the air layer causing droughts on earth.

Before the industrial revolution, seasons and temperatures appeared balanced, normal rain occurred that absorbed CO_2 back to earth. The greenhouse layer has now increased, it warms the ozone layer and evaporates the cloud causing more droughts. More droughts are occurring on earth, more ice caps are melting and sea levels are rising. These signals show that the CO_2 emissions have increased and global warming is occurring at a critical pace. These indicate that emissions of greenhouse gases have increased at alarming rate.

This is the reason some scientists have called on industrialized nations to reduce man-made carbon dioxide emissions in order to stop the melting of ice caps from the north pole and likewise prevent the occurrence of massive floods. At the same time, the heated atmosphere imbalances the equilibrium between cold and warm air resulting in the powerful rush of turbulent air, it creates an entropy of powerful winds that circulate and amass together on land toppling down multitudes of structures and homes. The rush of cold air collides with the warm air creating lightning that cause wildfires on earth. CO_2 accumulations in the atmosphere warm up the earth's atmosphere creating droughts and the rise of sea water levels. Scientists around the world conclude similar cycles from season to season. There are different variations of climates change depending on the position the clouds and the mass of the CO_2 floating above the earth as the earth daily moves around its axis and around the sun.

The same cycles constantly repeat itself, more warm air and cold air generated by the greenhouse layer season after season. The warmed layer of CO_2 causes drought, air turbulence, imbalance of cold air and hot air which results in the creation of powerful circulating winds sweeping water on land and bringing on floods on earth. CO_2 that mix with rain, falls in the ocean, contaminates sea water and makes water more acidic that impacts marine life. Global warming devastates agriculture on earth, impacts ocean marine life, resulting in enormous reduction of fish as well as dries up food products such as fruits, vegetables and grains that grows on earth. Global warming significantly reduces food products man needs for consumption.

There are now more fires and drought in summer. There are more floods, storms and greater heights of snow rising in winter. Climates balance are tipping to higher temperatures during summer and lower temperatures in winter beyond normal conditions. The effects of climate change poses greater risks than conflicts and terror groups in the world, like Katrina in Louisiana, USA and the Haiyan typhoon catastrophe in the central part of the Philippines, it wiped out hundreds of miles of communities that were swept out into the sea. Droughts and wildfires in California, Washington State, Nevada and other states extend thousands of square miles, destroying multitudes of homes and infrastructures in communities, including killing individuals and firemen along the fire lines.

Up until now, intents of governments to reduce greenhouse emissions have been made, but little has been done to sizably reduce carbon emissions. Alternative energy like solar, has not gone fast enough to replace fossil fuel production. Traffic vehicles have not been replaced by new green batteries. No one knows how long these alternative energy replacements are going to occur, it may take many more years. Meantime, more floods and storms are occurring in greater magnitude and strength and each time many lives are lost and billions of dollars in infrastructures and homes destroyed.

The application of Maharishi's natural law principle of invincibility and the cooling mechanism displayed by the human mind has been observed when large groups quiet their minds in large groups together. Maharishi's discovery of the Unified Field provides an instant and immediate solution to significantly reduce or eliminate CO_2 in the atmosphere and balance the weather. Evidence already exists displayed by many events Maharishi has called for during his lifetime. Maharishi referred to the finest particle of nature as the "sap" of the universe, like the invisible "sap" that penetrates the root of the tree. It nourishes the whole tree. Maharishi said that the Unified Field is the "sap" of the universe that is enlivened within the human mind at the level of silence and nourishes all creation. The Unified

Field is made up of the finest particles of nature, finer than oxygen and hydrogen in the air and far finer than the carbon particles. These finest particles from the Unified Field can penetrate the magneto particles around the earth's magnetic field and it is finer than the graviton particles hovering within the earth's gravitational field. The Unified Field can break thru the magnetic and gravitational field and thus can push the trapped the CO_2 layer and effectively disperse CO_2 into space above and around the earth. Meanwhile the Unified Field can instantly cool the air in the atmosphere during summer and then fall on earth as rain. The cooling character of the Unified Field was discovered to occur within the human mind in large groups together at the level silence. Maharishi's Transcendental Meditation quiets the mind in an easy and effortless manner.

Without a name available, I referred to the finest particles, as the ASu particle, the anima solo particle that nourishes and animates the whole universe and creates the Unified Field. Like Maharishi similarly referred to it as the "sap" of the universe that nourishes all creation, it significantly cools the atmosphere and turns H_2O into rain. When H_2O is cooled, it changes state to liquid form as water and at the coolest freezing point, it turns to solid ice or falls on earth as snow.

The CO_2 particle is not finer than the magneto or graviton particles to be able to penetrate the earth's magnetic or gravitational field that is why it is trapped beneath in between. It cannot break the air or cloud layer as well because it is heavier than air. It gathers up heat and evaporates the cloud, once air evaporates, the ozone layer is depleted and results in droughts on earth. The Unified Field particles push the CO_2 out thru to the magnetic and the gravitational field of the earth and they can be dispersed in space solving global warming. The ASu particle possess the natural cooling character of the Unified Field, cools the clouds and fall as rain. It is the natural character of air when cooled, it turns to water and falls as rain on earth. Experimental evidence displayed the cooling effect of the ASu particles lingering within the human mind at the level of silence in large groups together and illustrated by the instant relief of droughts by the application of the Unified Field.

I created the effect in small scale with just 400 students in my second TM school, rain fell around the school area and cooled the summer day in July 1986. The same phenomenon occurred, rain fell around the area in my fourth TM school with 500 students in 1994. The same rain phenomenon happened with 400 workers on the job building the thirty storey tower of invincibility also in 1994, not once but many times. Then, I was a young structural engineer who wanted to validate what Maharishi has said, "Happiness in man's birth right, not suffering. Only a new seed will bring a new crop." I witnessed the rain phenomenon many times, so I can confidently write about it and confirm without any hesitation that Maharishi's technology of the Unified Field truly works. I witnessed many times, the cooling effect of the Unified Field particles lingering within the human mind and the consequential rain phenomenon that was produced each time the Unified Field principle was applied in large groups together.

I was not alone in verifying the truth Maharishi has foretold. One of the many cooling event was displayed in the Global Taste of Utopia organized by Maharishi from December 16, 1983 to January 6, 1984 described in Chapter 6. There was a prevailing drought in Brazil. Maharishi promised rain would fall in Brazil during the Global Taste of Utopia and rain did fall as Maharishi foretold.

In Mozambique, Africa, in 1994, as soon as the Unified Field principle was applied by large groups of the military in Mozambique, Africa rain fell and relieved the drought in Mozambique that has prevailed nearly twenty years, at the same time the two decades of civil war that prevailed in the country ended. Maharishi was a physicist and found the cooling effect of the finest elementary particles of nature found to linger within the human mind in large groups together. Many events followed the same trends of weather balancing.

Maharishi has brought to light the Unified Field, it can now remove all suffering on earth. It can balance climates, it effectively counters terror groups, conflicts, and war, it makes economy rise and creates peace to prevail on earth. These are all contained in Maharishi's natural law principle of invincibility displayed by the discovery Unified Field. "It is like a tree," Maharishi said. "The sap" penetrates all aspect of the tree. Then the leaves turn green, the flowers blossom and bears fruit." In man, the Unified Field nourishes every aspect of life, mind, body, behavior and his environment blossom with positive trends. No negativity can exist within the Unified Field found enlivened within the human mind in large groups together.

Observation is the significant tool of science. If experiments done several times and the same results are observed each time, it becomes a truth. The cooling experiment has been done more than fifty times and the same cooling results occurred. This proves that a cooling solution to global warming has finally been found, just by gathering large groups of Maharishi's Transcendental Meditation and Yogic flying participants in one place together.

Droughts, storms and floods never had a solution, neither did volcano eruptions, earthquakes and tornadoes. The Unified Field is a recent find brought to light by Maharishi Mahesh Yogi. Since then, observations were done evidencing that the human mind has the capability to control his environment in large groups together, at the level of silence. The technology of the Unified Field can all together eliminate human suffering in our generation on a permanent basis.

Maharishi has described the ASu particles at the level of the Unified Field as the "sap" of the universe. Like a tree, the "sap" seeps from the root and nourishes all aspect of the tree. In the same manner, the Unified Field provides the same nourishment to the individual, community, country and the world. To the individual, it nourishes tired and sick cells, rejuvenates and restores aging cells and makes man healthy and promotes longevity. (See scientific researches on TM at wwwtm.org). In the community, it reduce crimes and diseases, for the country, it makes economy rise; for the world, it counters terror and makes conflicts and war end and finally, cools the atmosphere, balance climates and relieves global warming. Refer to Chapter 6, the Global Taste of Utopia, a global experiment on the assembly of the 7,000 TM and Yogic flying participants who gathered in groups together from December 19, 1983 to January 6, 1984 in Fairfield, Iowa, USA. The 7,000 is the square root of one percent of the population of the world then in 1983. It displayed the positivity produced around the world and drought stricken Brazil was relieved by the instant rain that occurred. The root of all creation, including the earth, man's abode, its forests and ocean habitat takes nourishment from the Unified Field of natural law. The ASu particles at the Unified Field contain the finest particle of nature that has the character and qualities Maharishi has enumerated as follows;

1. All possibilities
2. Bliss
3. Unbounded
4. Self-sufficiency
5. Integrating
6. Self-referral
7. Invincibility
8. Perfect balance
9. Fully awake
10. Freedom
11. Simplicity

12. Unmanifest
13. Infinite dynamism
14. Infinite silence
15. Pure knowledge
16. Infinite organizing power
17. Perfect orderliness
18. Infinite creativity
19. Evolutionary
20. Purifying
21. Nourishing
22. Infinite correlation
23. Harmonizing
24. Immortality
25. Total potential of natural law
26. Field of fulfillment

Maharishi described it like a master particle containing the nourishing aspect of natural law. Perfect balance refers to climate change. If it is too cold, it provides nourishing warmth. If it is too warm, it provides cooling balance. The human body posses a natural temperature sensor mechanism, when warmed, it activates the sweat glands and water oozes from the human skin. This illustrates that the human body posses a natural built in cooling mechanism, further illustrates the natural character of water, when cooled, turns to liquid water, When further cooled at zero degrees centigrade, water turns solid ice, when heated 100 degrees centigrade turns to gas. These are the three states of H_20, liquid water, solid ice and air. Its constitution, H_2O remains the same in its three states, liquid, matter and gas.

With the practice of TM and Yogic flying, man can activate the cooling mechanism built within him, he can also cool his own planet earth at the silent level of the human mind in large groups together, because his constitution is the same as the constitution of the universe and of the earth the at the level of the Unified Field, the field of the finest elementary particles of nature that posses the natural temperature sensor mechanism built within man that can cool global warming in an instant! Human sensor mechanisms have already been developed on earth, like a closed door opens when a man comes near, or a tablet unfolds a whole lot of information with just a touch of a human finger!

I included that it is also the field of fulfillment of desire in large groups together, thinking the same desire happens, also described in Maharishi Yagya program. Maharishi has brought to light these amazing character and qualities of the finest particle of nature also contained and enlivened within the human mind in large groups together.

Maharishi has said that in large groups together, man becomes master of natural law at the level of silence. Man has to find and draw from that deep silence thru Transcendental Meditation that Maharishi has provided man, a mental technique to draw down and quiet the mind thru the practice of Transcendental Meditation. Maharishi has unfolded the means and technique to balance man's environment for humanity to enjoy heaven on earth.

Large amounts of properties and infrastructures are wiped out and destroyed by floods, storms and wildfires constantly occurring on earth. Food products are significantly reduced by droughts. Many lives unnecessarily lost by natural disasters once thought to be an act of God and considered impossible to solve. A solution has finally become available and Maharishi has predicted that heaven on earth can be created. It is a reality within reach by humanity to enjoy. Man has the capability to create it.

Global warming is man-made. Man, in large groups together is capable to prevent and undo the harm it creates. The technology of the Unified Field has now been discovered. It holds the invincibility character and behavior of the finest elementary particles of nature also found to occur within the human mind in large groups together that can stop global warming at less cost, in contrast to the budget proposed by world climate agencies in the amount of $10 billion dollars or the $100 billion dollars by year 2030. If military groups are used like that once done in Mozambique, Africa, the cost is minimum as the TM and Yogic flying program is done twice a day in the morning and afternoon, before and after work.

At one time, Maharishi requested for $1 billion dollars donation from governments in the world to finance a group equal to the square root of one percent of the world's population to practice Maharishi's Unified Field program. No one government provided him a single dollar, so by himself, he took the responsibility before he passed away. Maharishi instead asked donations from TM practitioners from around the world and organized the project of the Vedic pundits (see Maharishi's Vedic pundits.org) at the Brahmasthan in India which began with just 1,000 participants that continues to this day. The 9,000 is the number of participants needed at this time and increases as the population of the world increases.

Since the founding of the Maharishi University of Management in Fairfield, Iowa, USA, groups of students and Fairfield residents create large numbers of TM and Yogic Flying participants. The group has been established for more than forty years, but until today, it has not formed the 9,000. More participants are needed to solve the urgent problem of the world, global warming. For smaller groups, the formula is the square root of one percent of the population of a community or a country which is just a couple of hundreds in some instances.

Meantime, droughts and massive fires have raged California, a huge flood has ravaged Northern Japan as it has ravaged Colorado, Texas and other states. Missouri and Illinois, USA was flooded during Christmas time in 2015, submerging homes and highways nearly thirteen feet deep. Louisiana has constant floods. Terror groups and conflicts still prevail in the world. Governments just need to do it and find the solution to the many problems prevailing on earth. Compelling proofs already exist that the Maharishi Effect[7] can transform the environment of the whole world, from negative trends to positivity on a permanent basis. Maharishi has foretold that heaven on earth can be created in our generation and the human mind has the capability to create it, just by creating large groups together in one place, the Unified Field is enlivened within the human mind at the level of silence.

---

[7] *See Chapter 4, The Maharishi Effect*

3

THE ANCIENT ORIGIN OF TM

The ancient Vedic civilization of India held the secret wisdom of meditation. Due to the long passage of time, it faded away and soon the knowledge was lost and forgotten. When the ancient Saraswati river dried up, the Vedic settlers along its river banks move away and became scattered. They sought other plains where water is abundant. A few Vedic masters, rishis they were called, they settled in the Himalayas where they carried on and kept the Vedic traditions from master to student from generation to generation and there in the Himalayas the tradition survived for many centuries.

Maharishi's master, Guru Dev born of a Brahmin family. The caste system survive in India describing the family social status, like the Brahmin caste belong to teachers and masters, the Ksahstriya warriors are administrators. Guru Dev went to the Himalayas at the age of nine and was sent back home by Swami Krishnananda, the prevailing master at that time. Five years later, at the age of fourteen, Guru Dev went back with a letter of parental approval and became the student and disciple of Swami Krishnananda Saraswati until the age of seventy when he became the Shankaracharya of Jotir Math in the Indian Himalayas. The seat of the Shankaracharya as left vacant for nearly one hundred fifty years. After repeated request, in 1941, Guru Dev accepted the position of the Shankaracharya representing the highest ancient religious order still prevailing in India today.

Maharishi, after graduating from Allahabad University in India with a physics degree, he sought out his master, Guru Dev in the forest of the Himalayas. Maharishi said of Guru Dev, "The first glimpse of him, I know he was the man I wanted to serve." Maharishi described the life of Guru Dev, "The greatest achievement of a saint is his life itself, the high edifice of realized Upanishadic living that develops from direct experience of reality. To understand that inner personality, one must approach such realized souls with open mind and try to visualize the great internal life that is the basis of their actual and real form of living." Maharishi became Guru Dev's secretary and stayed with Guru Dev for twelve years until Guru Dev passed away in 1953. After Guru Dev passed away, Maharishi retired in solitude in the caves of the Valley of the Saints in Uttar Kashi, deep in silence in the high mountains of the Himalayas.

There was another old man who would sit quietly beside him each day as Maharishi ventures outside of his cave watching down upon the city beyond. Maharishi whispers, "Rameshvaram" a city thriving below the Himalayas. One day, the old man said, "Why don't you just go to Rameshvaram, so you do not think of it so often and then come back."

In 1955, Maharishi came down to Rameshvaram and never went back to his mountain abode. It was a turning point in the history of mankind. Maharishi brought with him a new seed sprouting that contained the blue print of a new civilization for all mankind, a life beyond problems and suffering. Maharishi spoke of the wisdom of the Vedic masters taught to him by Guru Dev. This technique was given only to a few chosen disciples. Maharishi broke the tradition and taught it to many, nearly six million people who have learned the Transcendental Meditation technique around the world. He elaborated that his Transcendental Meditation can make man live a life of happiness and fulfillment and should be taught to everyone.

Maharishi recalled later, "Coming from the Himalayas, I had the thought, it was very fresh in my mind, that it is not necessary for man to suffer. All the literatures of India is full of beautiful exaltations. The Veda says, "O sons of Immortality," addressing all mankind as sons of immortality. The Veda says, "Of Anand, (bliss), all this is born, in Anand, all this is sustained, and to Anand all this is going. But where is the reality of this in the day to day life of the people?" The ancient Vedic literatures extolled the unseen, unmanifest and finest form of creation. Modern science called it the Unified Field.[8]

With this singular purpose, Maharishi went on to teach Transcendental Meditation to people in the world so that they can integrate bliss consciousness in their daily life. With the experience of bliss experienced by many who practiced Transcendental Meditation, it gradually spread throughout the world. Maharishi Mahesh Yogi's fifty years of history spanned his teaching of Transcendental Meditation. During his lifetime, he brought to light the most significant knowledge on how to remove problems and suffering that have become a daily encounter in the life of men. Many philosophers of ancient past have sought out mechanics which can improve the well being of humanity. None has compared to the effect of the Transcendental Meditation and the Maharishi Effect[9] that brought to light the transformation of man and his environment which future generations can apply. Maharishi is considered the foremost teacher

---

[8] *Chapter 8, The Discovery of the Unified Field*

[9] *Chapter 4, The Maharishi Effect*

in unfolding human consciousness as the basis to the solution that can alleviate the many maladies that plagued mankind for centuries. Maharishi is considered the greatest teacher in human transformation in our modern times.

Maharishi has gathered the old, ancient and scattered Vedic literatures that described happiness is the real and true birthright of man. It transformed thousands of years of struggle, suffering and problems that characterized the known history of mankind. The ancient Vedic literatures contained the knowledge that man can live in accord with natural law governing the whole universe in harmony and perfect order. The identity of man refers to his cosmic character in tuned with natural law. The Vedic literatures described the cosmic nature of man and his birthright to happiness. The knowledge of transcendence was lot in time and man started to suffer and his birth right was forgotten and shadowed into a long, long history of obscurity.

Transcendental Meditation originated from the ancient Vedic civilization extolled in the ancient Vedic literatures from that long forgotten age which described man can live in the same harmony and order as all the planets in our galaxy that move in perfect harmony and order. Man is part and parcel of the universe and can live in that same order and harmony and need not have problems and suffer.

In the ancient Vedic civilization, the Vedas described the reign of King Raam, where no one suffered and everyone lived in affluence and peace. People lived in accord with natural law. During that period of ancient history, the norms of the ancient Vedic tradition prevailed. Maharishi brought back the technique that can bring the same affluence and fulfillment to everyone in the modern world thru the practice of Transcendental Meditation and Yogic flying in large groups together. Transcendental Meditation allows the human mind to access that level of natural law at the most silent state of the human mind.

Maharishi's achievement surpassed teachings from many ancient masters. Maharishi compiled the ancient Vedic literatures like a manual for man to live life in fulfillment and happiness. He has gathered the ancient Vedic literatures and transformed it into Vedic science that can be applied in our modern times. Maharishi was able to expand the knowledge of natural law contained in his Vedic Science of consciousness thru 192 countries throughout the world. Maharishi's Vedic science is now available at the Maharishi University of Management in Fairfield, Iowa, USA.

In a period of fifty years, big nations like the USA, Great Britain, China, France, Germany, Japan and even Russia, hundreds of its citizens have learned Transcendental Meditation. Nearly one million in the USA alone have learned TM including several celebrities and many men of fortunes. From all over continents including Africa, Europe, Asia and most cities in the world, big or small country like the Philippines, a Transcendental Meditation center has been established. Six million individuals have learned Maharishi's Transcendental Meditation since Maharishi brought it out into the world. TM has gone main stream in the USA and nearly twenty thousand[10] individuals sign up and learn TM every month including many Wall Street participants. Based on personal experiences and hundreds of scientific validations from distinguished scientists from renowned universities worldwide, Transcendental Meditation is a system that unfolds the most significant and enduring human transformation before Maharishi came down from the Himalayas in 1955.

The popularity of Maharishi began in the 1960's. The Beatles learned Transcendental Meditation from Maharishi and they followed and toured with him. The Beatles rose from obscurity to become the most popular and most sought after rock singers in the pop culture that was fast rising at that time. It was the most extra ordinary period of creativity and success for the Beatles. In a recent interview with Paul McCartney, one of the surviving Beatles, he commented that TM was a gift from Maharishi when they were seeking something at that time to stabilize them. Transcendental Meditation took the Beatles off from drugs. Their quote, that they were, "more famous than Jesus Christ," created worldwide controversy among many Christians during the peak of their rock popularity that begun in the early 1960's. Maharishi similarly rose to unprecedented fame.

It was told that Maharishi was born in January 12, 1917 in a small village of Chichli, India. The social culture of India categorize family status as the Brahmin caste, consisting of teachers and scholars; Ksahtriyas, the class of warriors and administrators; Vaishyas, the caste of farmers and traders; and Shudras, the worker's caste. After his death, Guru Dev transferred authority of the Shankaracharya of Jotir Math to a Brahmin required by its code. Maharishi's family belonged to the second caste, a Kshatriya, a warrior caste Hindu family.

---

[10] *November 2015 announcement by John Hagelin, PhD, National Director, Maharishi Foundation, USA*

Maharishi studied physics as a student, at the Allahabad University in India and earned a master's degree in 1942. After college he worked for a while and then went to the Himalayas to search for a master who can teach him the secrets of the Vedic tradition that was hidden from the world and held by only a few masters, called Rishis in the Himalayas.

Many in population recall the prosperity and peace of its ancient Vedic history and believed that the Himalayas held and kept the secrets of the ancient Vedic literatures expounded by wise men, masters of the ancient tradition of knowledge, for these reason they were much sought after. The Veda and Vedic literatures originated from the ancient Vedic civilization of India and the ancient Vedic norms were transferred from one family to the next family so that the ancient tradition will go on from generation to generation.

Serving a master was once part of the ancient Vedic culture that still holds on today in modern India. Masters are greatly revered and students were highly exalted for following the ancient tradition of masters which the Vedas states, "He who serves a master holds the total knowledge of life."

The secret tradition was sought after by seekers of knowledge from many distant places. Maharishi had acquired this credibility, having stayed in the Himalayas for some time and served a master, Guru Dev who was exalted in high esteem by the Indian population, so many people listened to Maharishi.

By 1957, Maharishi established several Transcendental Meditation centers in India and spoke of regenerating the world. "We believe men are in need of something that would develop their latent faculties, help them to rise above the cares and anxieties of life to a high pedestal of peace and happiness and increased efficiency in the field of action. Such men do not belong to any one particular province or land, and they are not the followers of any one religion or philosophy. They are the citizens of the universe and are scattered all over the world, following different religions. For them all is meant this Spiritual Regeneration Movement," Maharishi said.[11]

The prolonged clapping from the vast audience inspired Maharishi, by then thousands have learned Transcendental Meditation

---

[11] *Maharishi, 1 January 1958, Madras, India*

and large crowds listened and followed him. Maharishi said, "When this is the response from Mother Nature on the thought of spiritually regenerating the world through this practice, we'll inaugurate a world movement, we'll inaugurate Spiritual Regeneration Movement here tomorrow evening." Eventually, one man invited Maharishi to come to America and paid for his plane ticket. Apparently this started his global tour around the world. Many press people met him and gathered at the Los Angeles airport to interview him. Some wrote, "This guru from the Himalayas, he comes to teach a meditation that can make you sleep." Transcendental Meditation can relieve insomnia. "He has no money. He holds his worldly possession in one hand." Maharishi had little personal possession that he carried in one small bag. Maharishi wore a piece of ivory-white long silk tunic attire and a pair of plain wooden sandals, characteristic of a Himalayan guru. Maharishi became an instant celebrity in America. He spoke publicly the benefits of Transcendental Meditation. And soon, TM became more than a popular form of relaxation technique in the USA.

In June 13, 1959, the Spiritual Regeneration Movement established its chapter in Los Angeles where it held regular weekly meetings and held courses with Maharishi and in every way "watered the root" expanding the population of Transcendental Meditation several thousands. In the early 1960's, many celebrities followed Maharishi. It was a time when many young people prevalent at that time roamed the country, they were called the flower group, they also followed Maharishi.

In 1970, scientists participated in documenting the benefits of Transcendental Meditation, and the first scientific research on Transcendental Meditation was published. More than 600 research studies were conducted and published from more than 200 prestigious universities from many countries.

The positive results from the scientific researches caused rapid growth of Transcendental Meditation and thousands from around the world learned the TM technique. Maharishi's popularity grew as he was invited to speak at several TV shows and was constantly interviewed by the press and media groups. The benefits of TM spread rapidly by word of mouth made by many positive experiences from many celebrities and ordinary people who practiced Transcendental Meditation.

Maharishi appeared twice on the then popular Merv Griffin show that resulted in tens of thousands of new TM practitioners. In 1975, he was at the front cover of The Time magazine. He appeared on CNN's Larry King's Live TV show several times.

Maharishi's world tour included Myammar, Singapore, Australia and Thailand in the Far East. He traveled to most countries in Europe including Norway, Finland and Switzerland. He visited Canada, Brazil and almost all continents. Maharishi traveled around the world seven times in thirty one days.

TM centers were opened in major cities worldwide. Mentmore Castle in England became a TM center. Livingstone Manor in New York also became one among many TM centers in North America. By then, in the early 1970, Maharishi has created a world movement that held the teachings of TM that still persists today.

Then in 1974, the TM movement of the USA purchased the old Parson College in Fairfield, Iowa, USA. It was named after Maharishi, previously the Maharishi International University, now became the Maharishi University of Management. Fairfield has now become a popular TM town in the USA. Majority of residents of Fairfield practice Transcendental Meditation. Fairfield has become a destination for many out-of-town and foreign visitors who come to join the group practice of TM and the advance TM-Sidhi program at the Golden Dome of Pure Knowledge located at the campus of the Maharishi University of Management. Maharishi appointed Dr. Bevan Morris, as the President of the university, the position he held for forty years. He has now retired. Dr. John Hagelin took over.

In 1974, it was observed in four cities in the United States, when the number of people participating in the Transcendental Meditation program reached one percent of the city's population, the trend of rising crime rate and low economic index was reversed indicating increased coherence and harmony in the life of the whole city. Scientists called the increase in harmony in collective consciousness of the whole society, the Maharishi Effect in honor of Maharishi who predicted it as early as 1960 about the time he started his world wide movement. Hundreds of cities in the US followed the same trend.

In 1977, Maharishi brought out the advance TM-Sidhi or the Yogic flying program that was found to be one hundred times more powerful than the basic TM. Maharishi started to train teachers of Transcendental Meditation. "I will multiply myself," he said. There are now some 40,000 teachers scattered around the world.

As a physicist, Maharishi correlated the Maharishi Effect to the Miessner effect in physics which illustrated similar behavior of the magnetic field effect that increases when more magnets are joined together, they produce a much greater magnetic field invincible from negative outside interference.

There are several benefits derived from the TM practice. These benefits have been tested and have been published in many prestigious journals. Most sought after benefits of TM are for health reasons. TM lowers blood pressure, relieve heart disease and reduce many other chronic diseases. It reduces anxiety and depression. Most people learn TM for self improvement and well being. The Maharishi Effect creates a broader and wider effect. Climates balance. War and conflicts end. Positive trends prevail in the environment.

As a single mother, I pursued TM to provide a better life for my children. I have completed my duty to my children as they have all grown and now have their own families. All love is directed towards the self, as described in the ancient Vedic book, the Upanishad. A fully enlightened man directs love towards the good of humanity. The love of a leader should be towards the good of each citizen, like the role of a mother exploring the best means to provide a better life for them. Transcendental Meditation provides it.

4

THE MAHARISHI EFFECT

In 1974, the Maharishi Effect was observed when one percent of the population practiced the Transcendental Meditation technique in eleven cities in the United States. It caused the reversal of the general trend of rising crime. One percent of the population was sufficient to neutralize all negativities in the surrounding environment. The phenomenon of rising coherence and harmony occurring in participating cities was called by scientists, the Maharishi Effect in honor of Maharishi, founder of Transcendental Meditation.

Maharishi explained the effect by the enlivenment of the Unified Field of natural law found in the transcendental field of pure consciousness located within the most silent level of the human mind. The phenomenon is a realization of the promise Maharishi made in Madras, India in 1957 that would happen in society during the early days of Maharishi's TM movement. His prediction would be a rise in positivity in society by just one percent of the population of a city or country practicing his Transcendental Meditation.

It is a natural 'field effect' described in physics produced by the elementary particles of nature,[12] like the magneto particles create the magnetic field or the electron particles create the electric field pervading around in the universe. The Unified Field is the 'field effect' created by the finest elementary particle of nature lingering within the human mind at the level silence in large groups together. For my reference, I referred to it as the AS_U particle, the anima solo particle that animates the whole universe. Maharishi referred to it as the "sap" of the universe that creates "from itself, by itself, and within itself," that is why I referred to it as a solo particle, the finest primal particle of all of creation where all matter forms spring from. It is the seed form of natural law.

Natural Law is responsible for the origin and evolution of everything in the universe. The unmanifest, non-material field of Natural Law is characterized by perfect orderliness, infinite creativity, unlimited energy and pure intelligence. Natural law is the will of God. In Vedic science, this field is called Brahm, the field of peace and bliss. It is like the sap within a tree which gives rise to, and

[12] *See Chapter 7, The Science Behind Climate Change*

nourishes all aspects of the tree. Physicists call this field, the Unified Field of Natural Law. Since Brahm, or the Unified Field is all-pervading, it is also enlivened within the human mind at the level of silence. The simplest form of human awareness - where the mind is completely silent yet fully alert, it is the home of Natural Law within man. This vacuum state or ground state of the mind is called pure consciousness, the Self. So each person already has that field of peace and bliss within that only needs to be drawn thru the practice of Transcendental Meditation.

Maharishi Mahesh Yogi says, "It's as if we live in a house which has a vast treasury in one of its rooms. Only we've forgotten about it. So, instead of living a life of royalty, we go about in poverty. Transcendental Meditation is a simple, effortless procedure for locating this inner treasury, by allowing the mind to dive within, and experience pure consciousness."

Maharishi said, "The Self of everyone, the Atma of everyone, the transcendental field of reality of everyone, is the same in everyone. Whether the body calls itself an American, German, Indian or Chinese, it doesn't matter. The house may be of anything, but the 'in-dweller' of the house is the same: the eternal, immortal, blissful Self, the Brahm. The whole of Vedic literature is full of knowledge about it."

The discovery of modern science of the Unified Field defined the meaning and interpretation of the Vedic science referring to the Self or Brahm contained in the Vedic literature where Transcendental Meditation and Yogic flying came from.

In 1977, Maharishi introduced the more advanced TM-Sidhi or Yogic Flying program, a more powerful effect that is one hundred times more powerful than the basic Transcendental Meditation technique. It would consist of only the square root of one percent of the population practicing the advance technique of TM-Sidhi or Yogic flying. This much reduced requirement, in most cases is just a few hundred participants which could bring life in society in accord with natural law.

Negativity cannot exist in the Unified Field, because it is the purest field in nature, only the positive aspect of natural law can exist. Life and the whole surrounding environment will be protected by the invincibility principle of natural law[13] characterized by a problem free

---

[13] *See Chapter 14, Invincibility Principle of Natural Law*

environment. Climate will balance, wars and conflicts will end, prosperity will flourish and all negativity will fade away.

It is like no impurity can exist in the magnetic field or the electric field will burn any impurity that comes within its field. Man and his environment will be protected by the invincible Unified Field. It's a natural phenomenon of nature further explained in Chapter 14, The Invincibility Principle of Natural Law.

It is a universal principle in nature that coherent systems possess the natural ability to repel external interference while incoherent systems are easily penetrated by disorder from the outside. It is like the laser effect in photon technology. More condensed heat particles produce more heat than only produced by one heat particle. One hydrogen particle acting alone has an insignificant effect, than groups of them contained together can produce a spectacular big blast.

The first major test of the Maharishi Effect as predicted by Maharishi took place in 1978 during Maharishi's Global Ideal Society Campaign in 108 countries. Crime rate was reduced everywhere. The global research demonstrated a new formula of the square root of one percent of the population practicing TM and the TM-Sidhi program or Yogic flying morning and afternoon, twice a day in one place together is sufficient to neutralize negative tendencies and promotes the growth of positive trends in the whole population. This discovery was repeatedly verified on the city, provincial and national levels.

When Maharishi promoted Yogic flying on a global level, with the square root of one percent of the whole world population participating, the 'field effect' was witnessed throughout the world strengthening positive and harmonious trends in world consciousness. The greatest evidence of this effect was the end of the enmity between the two super powers, the cold war between the Soviet Union and the United States ended with a friendly handshake in 1988.

Further on, the Global Taste of Utopia[14] was conducted in December 1983 in Fairfield, Iowa, USA with the square root of one percent of the population of the world at that time in 1983 equal to 7,000 participating at the Golden Domes of Pure Knowledge located at the campus of the Maharishi University of Management, it illustrated to the world that heaven on earth can be established that was confirmed and tested repeatedly proving its indisputable scientific basis.

---

[14] *See Chapter 6, the Global Taste of Utopia*

A scientific theory is proven true when the results are repeatedly observed to be the same after each repetition, then it becomes a fact. The events were repeated fifty times illustrating similar results of positivity in the environment including balancing climates. Maharishi predicted that rain would fall in Brazil during the Global Taste of Utopia in 1983, and the rain did fall. The group effect brought rain to the drought stricken Brazil. The Maharishi Effect balanced the warm climate. In Mozambique in 1994, during the term of then President Chisano of Mozambique, the drought that lasted twenty five years ended when rain fell and nourished the dried earth as soon as the military of Mozambique started practicing TM and Yogic flying together.

I attempted to validate the Maharishi Effect, it resulted in similar positivity rising in the surroundings. When my second TM school started to meditate, it rained in July in 1986 after the students started to meditate. The same thing happened in my fourth TM school in 1994, rain fell the moment the students closed their eyes in meditation. balancing the hot July climate. In the same year, in 1994, when four hundred workers started to meditate, the wild rain stopped during the rainy season and the floods receded in the flood zones surrounding environment. Four hundred workers created the Unified field that balanced the climate. I attached my story in the following chapters.

Many carefully controlled experiments on the Maharishi Effect have appeared in leading scientific journals such as the Journal of Conflict Resolution, Journal of Crime and Justice, Social Indicators Research, and Journal of Mind and Behavior.

These studies have utilized the most advanced and rigorous research designs and statistical methodologies (time series impact assessment analysis, a special case of Box-Jenkins transfer function analysis) to precisely evaluate the effect of large coherence creating groups on standard sociological measures of the quality of life in cities, provinces, nations, and the world.

These studies have rigorously demonstrated the power of the Maharishi Effect to a degree of certainty that is unparalleled in the sociological sciences, and even in the physical sciences. Thus the Maharishi Effect has been more extensively documented and thoroughly established than any other phenomenon in the field of scientific research. The Maharishi Effect in itself proves the existence of the Unified Field of Natural Law and man's ability to operate from this level. The phenomenon of the Maharishi Effect (like the phenomenon of the Meissner Effect in Physics) discovered by scientists has repeatedly verified that coherence in collective consciousness and positivity and harmony in national consciousness is produced by large groups practice of Maharishi's Transcendental Meditation and Yogic flying in one place together.

Maharishi said, "All good to everyone and non-good to no one. With this, the two thousand years old gospel of suffering is soon coming to an end and a new civilization, Vedic civilization of enlightenment and fulfillment is dawning."

As the Maharishi Effect is purifying world consciousness, old principles that were guiding life in the age of ignorance are being replaced by new principles that will guide life in the world in the dawning Age of Enlightenment.

Maharishi repeatedly emphasized the significance of the knowledge of the ancient Vedic civilization where his teachings, TM and Yogic flying originated. The ancient Vedic civilization prospered because they chanted the sounds of the Veda generation after generation which are sounds of natural law. They lived with the ancient Vedic norms that guided life in accord with natural law that is problem free.

The Maharishi Effect was described in the Vedas, it says,

"Go together, speak together. Know your minds to be functioning together from a common source. In the same manner as the impulses of creative intelligence in the beginning remain to be together near the source. Integrated is the expression of knowledge."

"An assembly is significant in unity. United are their minds while full of desire. For those who make use of the integrated expression of knowledge, by virtue of unitedness and by means of that which remains to be united, perform action to generate wholeness of life. United be your purpose, harmonious be your feelings and collected be your minds in the same way as all the aspects of the Universe exist in togetherness, wholeness."

At the level of the Unified Field where the Maharishi Effect operates contains exponentially powerful effect when group minds desire together the same intent at the most silent level of the human mind, desires are fulfilled. Like in the electric field, the copper wire conducts heat, the human mind and nervous system conducts the Unified field for the fulfillment of peace, prosperity and happiness.

Maharishi compiled all the Vedic norms into one complete science, Vedic science now available at the Maharishi University of Management in Fairfield, Iowa.

Change in Crime Trend in Major U.S. Cities (Over 250,000 population)

1975–1976 (January–June), compared with 1971–1972 and the national average.

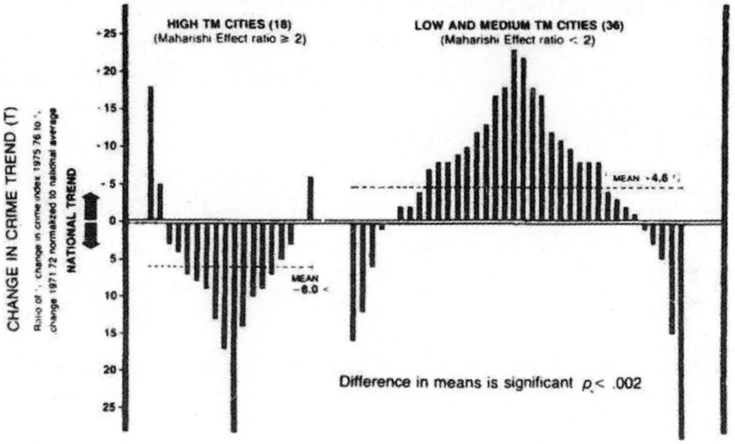

In 1976, the rising crime rate in the United States was brought to a standstill by the Maharishi Effect. Cities with a high density of practitioners of the Transcendental Meditation technique account for the reversal of crime trends experienced by the nation as a whole in the first half of 1976.

REPLICATION OF THE GLOBAL MAHARISHI EFFECT

Three large assemblies of experts in Maharishi's Transcendental Meditation and TM-Sidhi Program, including Yogic Flying, approaching or exceeding 7,000 in number (the square root of one per cent of the world's population at that time), were held from 1983 to 1985. These large World Peace Assemblies, held in Fairfield, Iowa, USA (17 December 1983 to 6 January 1984), the Hague, Holland (21 December 1984 to 13 January 1985), and Washington, DC, USA (July 1–10, 1985), were found by research to have the following results:

- Reduced International Conflict (Vol. 5: 411, Worldwide)
- Reduced Terrorism (Vol. 5: 411, Worldwide)
- Increased World Stock Prices (Vol. 5: 411, Worldwide)
- Increased Cooperation, Reduced Level of Conflict, Reduced War Fatalities and Injuries (Vol. 5: 410, Lebanon)

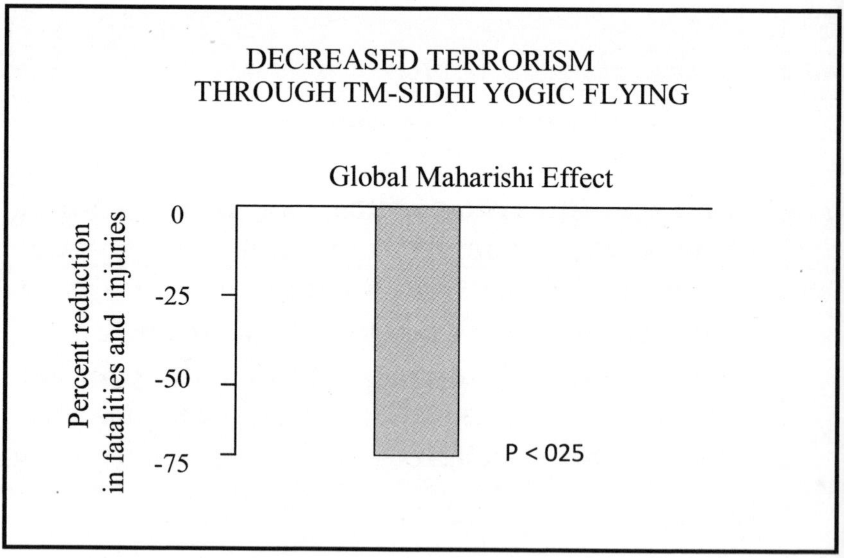

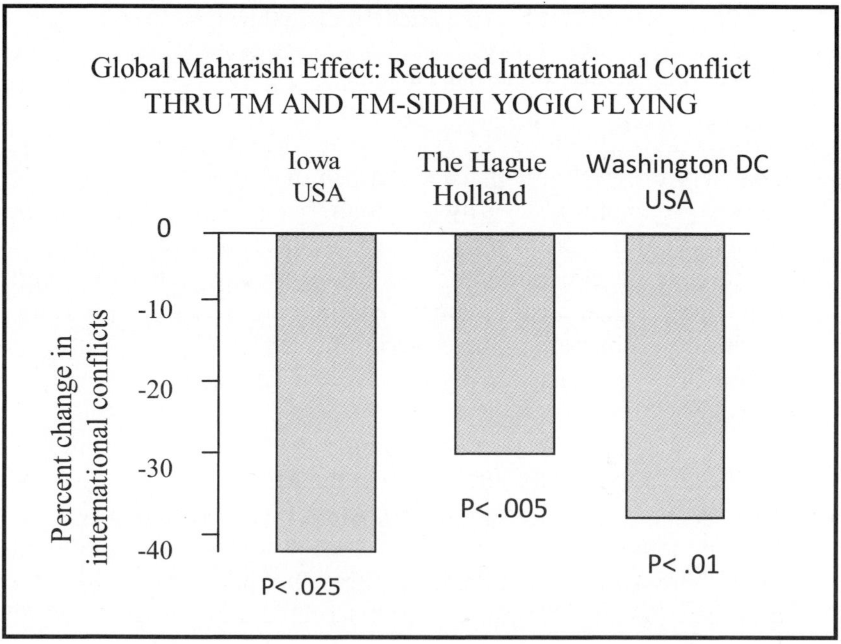

There was a significant decrease in fatalities and injuries due to international terrorism during the period of the three assemblies approaching or exceeding 7,000 experts in the Transcendental Meditation and TM-Sidhi program (the square root of one percent of the world's population at that time) as indicated by time series analysis of an independent data bank. **Reference:** Time series impact assessment analysis of reduced international conflict and terrorism: Effects of large assemblies of participants in the Transcendental Meditation and TM-Sidhi program. Paper presented at the American Political Science Association Annual Meeting, Atlanta , Georgia, August 1989

5

HAPPINESS IS THE BIRTH RIGHT OF MAN

I first met Maharishi in January 12, 1985. It was barely one month since the day I learned TM in December 12, 1984. After four weeks into meditation, I received a call, "You must go to Washington D.C, tomorrow. You will see Maharishi there."

"Tomorrow? Why so soon?" I replied very unsure of myself. "I still need to pack." It was an invitation to an international conference on education and Maharishi was expected to attend. At that time, I had also a small engineering school supporting my engineering company. There were other educators who were invited to attend as well. I hurriedly packed and left for the airport the next morning.

When I arrived in Washington D.C, I found that my luggage had been inadvertently left behind at the Manila airport. I had nothing else, only my small handbag. So, when I met Maharishi for the first time in the morning of January 12, 1985, I was wearing an old pair of faded blue jeans. Coincidentally, the following day was Maharishi's birthday. January 12 was always celebrated around the world giving homage to Maharishi who founded Transcendental Meditation. Every one dressed up for the occasion. I wore the only one attire I had on, an old denim blue jeans. The men wore suits and ties, the ladies, colorful Indian saris. During a photo session, someone placed a white shawl flowing over me covering my blue jeans as I was brought to sit beside Maharishi.

Maharishi spoke about the qualities of the mind at the level of silence. "The knower, the process of knowing and the known, Rishi, Devata and Chandas. The mind operates in three in one process, in one single point value, three in one," he said. How can the mind be a trinity in one? It was all Greek to me. Listening to Maharishi's discourses about meditation and the mind apparently seemed strange to me at that time. It just passed thru one of my ear and in the other leaving no impression. With jet lag and all, I just sat sleepy, uncaring about what he was talking about. All that interested me was feeling fine after meditation, beyond that, TM was just that then. I did TM regularly twice a day like a good new habit I have just started to get used to. It was only later that I began to understand the enormous and spectacular knowledge that is contained deep beyond the simple technique of Transcendental Meditation. I wondered why Maharishi was speaking and repeating the same topic over and over! I realized later the great significance of what he was talking about. The Unified Field and nature, the mind and nature, they all seemed meaningless to me at that time. Then, I was unaware of all the spectacular effects of TM, just the immediate relief and simple relaxation that I experience each time was good enough effect on me.

The most mind boggling phrase that woke me up was what Maharishi said, "Happiness is man's birth right, a life without suffering and problem-free world." It left an indelible mark in my heart that never went away, wondering in contrast to what my religion preached that suffering was part of life.

I was baptized a Catholic. Understandably, I have been born in a country where the overwhelming majority of the population are Catholics, at that time when approximately 80% of the Philippine population practiced the Catholic religion. I carried my Catholic religion by birth and I practiced it like everyone else. I went to church regularly on Sundays, prayed the rosary diligently every night, received communion and tithed a little each Sunday like all the other Catholics in the population did. Many religions profess suffering as part of life. Jesus Christ, the God of the Catholic religion suffered and died on the cross, thus immolated suffering and followers took suffering as part of daily life.

I was married a year after graduating from college. It did not last long. I became a single mother with three small daughters at a young age. I took whatever suffering in stride, and went thru some suffering bravely and did not complain. I took whatsoever suffering as part of life, like my religion dictated. There was no child support in a Third World country like the Philippines. I had to support my three small children the only way I knew, to work hard to provide the best possible life for all of them and myself.

I had no dreams of fame and fortune. I worked every single minute and had no time to attend any social gathering unless it was for business and I had to go. I wanted just to get on living a simple life, in a comfortable lifestyle with my children.

With a civil engineering degree I received from college, I initially worked alternating with the two prestigious engineering company in town, accepting the offer of a good salary and I took the higher salary bid each time. I started my own engineering company at the invitation of one of the clients of the big companies that I had worked for. A famous architect client at that time said, "You always come in early before eight o'clock and take all the messages before your boss arrive. Your other bosses always come late. I heard you work very hard and very good in your expertise. Why not put up your own company and I will support you," he offered. I could not refuse the old man's offer. He was one of the well-known architects at that time in the Philippines and owned the biggest architectural office, second only to then architect of the Philippine's First Lady, Imelda Marcos. Just a client with his business would be more than sufficient to sustain me, so it was not a difficult offer to accept. I was happy that soon, the clients of the other companies I had previously done work for followed and gave me work too. It was flattering, my hard work was paying off.

Of course, the companies I previously worked for were not happy with me for taking their clients. The president of one of the big company I used to work for wouldn't speak with me when we meet at the architect's center. It was my company that had previously provided the engineering design for the center when it was bought as an old church then restored to become a large hall for dining and function rooms where engineers and architects regularly meet and dine. Most of the meetings and affairs of the engineering, construction and architects group were held at this Architectural Center Club that my company participated in the engineering design that soon became a popularly frequented renovated building.

I worked hard to support my children, not because I wanted to excel in my engineering career, but by good fortune, my efforts paid off as I got many jobs from several clients. My peers fondly poked at me and called me many names, that wonder girl or something, whatever they called me then, as I worked toe to toe with the two big engineering companies that I had previously worked for. At that time the country was just starting to get developed in the late 1960's. There was a severe brain drain of engineers as many engineers including most of my classmates left for foreign countries seeking better income. It was a period of opportunity for my engineering company with just a handful of competitors.

After several decades, I nearly forgot all the projects my company had done then. When my eighteen year old grandson came over to California for a visit, He said, "Grandma, that church, shaped like a dome you designed, in a beautiful garden surrounded with a lake at the middle of the city, it has now become a popular beautiful landmark."

"Your company is among those listed at the lobby of the 22-storey Manila Medical Center too." He named a few more. Entry portals of erected buildings usually show names of engineers who participated in the design and construction. I was so surprised by his proud remarks. He was not yet born then. His mother must have told him. I did not take notice of the end products of my efforts. Then, I was just too busy working. After completing each project, it was enough to see my clients satisfied. I just moved on working day after day, sometimes tagging all my small kids along like a good mother, when I had to finish some urgent work on Sundays or holidays.

It was very difficult running an engineering and construction company and simultaneously running a small engineering school in a Third World country at the same time caring for my three small children. For a young mother, it was even more difficult. If you can imagine the difficulty for a single mother in the United States, with support for women-owned businesses, it was considerably more imposing a task working in a Third World country with no child support or any support for that matter. With the pressure of business and work, by the end of the day, I would be so tired and stressed out. That was before I learned TM. Many of my days would come to an end with a throbbing head ache.

One evening after arriving home tired from work, I received an unexpected call from Joey. He was one of the colleagues in the building industry. His family owned several buildings in the city. It happened that he was a practitioner of Transcendental Meditation.

He told me, "You must come to my place for dinner tonight. I have some people I want you to meet." I replied, "It's rather late, and I had a very busy day, I am too tired. Some other time, thank you anyway." Before I could put the telephone down, there was a knock at my door. It was Joey's driver. "I was instructed to pick you up," he said. I glanced at my watch. "It must be a very important dinner party to be going on at this time," I thought, as I hesitantly put my shoes back on and followed the driver to the car.

When I arrived at Joey's place, I was surprised to see just a handful of people in attendance. It seemed like I was the only guest present. There was a group of young men who were members of a small band playing some music. There were two other ladies present. I was told they were leaders of the local TM movement. The band started to play as I entered the room. I took a seat between Joey and the two other ladies present. For a little while, we quietly listened to the music played by the men in the small band. I glanced at each one of the musicians out of the corners of my eyes so Joey wouldn't notice my eyes roving.

Most of the men were tall and fair. They must be Americans, I assumed. "This party is just for you," Joey said, as food was served from a nearby table. After the band stopped playing, Joey stood up and nodded for each of the men to come forward.

Joey said, "I want you to meet the Unified Field band. They belong to the Transcendental Meditation group. They have gathered here in the country from different parts of the world. Maharishi had sent them here." I extended a hand shake as each one greeted me.

"Who sent them here from all around the world? Maharishi who?" I curiously asked. "I've never heard of him," I whispered leaning towards Joey. He explained to me who Maharishi was and why they were in the country. "They are here, about one thousand two hundred of them to quell the turbulence in our country," he said. At that point, I was too tired and sleepy to understand the narratives he was talking about. I knew there were a lot of people rallying and burning in effigy of then President Marcos in the streets of Manila. It was surely a difficult time for our country. The people were protesting in the streets. The opposition groups gathered in the streets in multitudes rallying against then Marcos, the President of the ruling regime at that time. It was a frightening time to be moving about in the city. People were marching with raised fists, angrily yelling, "Down with the dictator! Down with Marcos!" There were multitudes of people shouting in the streets! Frightful burst of noises in the streets was a daily scenario.

I remember my daughters clinging to me fearfully as we watched the noisy rallies from the fifth floor unit of our residence at the Ritz Tower condominium that I owned at that time in downtown Makati area. After the fall of Imelda's husband, Marcos, the former First Lady stayed in one of the units above mine at the same tower. I sent her my first book on TM. She did not reply.

The noises and shouts of the people in the streets beneath, down below our residence scared my children. I tried to put on a face of confidence as I was more scared inside than they were, as I had their future and their welfare as well as my own to think about.

There was so much turbulence in the streets! I couldn't help but wonder what good event the future would bring us all. I just had to work double time to tackle problems occurring in my office. Looking back, I wondered how I managed to survive. It must be my daily practice of TM that gave me a clear mind and strength to handle the turbulent times in the country. I can imagine the ordeal of mothers in other countries that have the same people protests and appraising. We now witness similar events in the Middle East, Iraq, Egypt, Libya, Afghanistan, Ukraine, Yemen and parts of Africa!

In Syria, scenes are truly tragic with many homes destroyed and multitudes of refugees, with helpless women with children! These are common pathetic scenes, mothers and children hugging closely together in fear of their lives and unsure of their future! Families and children gassed by chemical weapons in Syria, how frightening indeed! How tragic families fleeing away from the war and along the way, some die at sea, like the little boy that drowned and died, swept away at a nearby shore. It was too much to bear.

At that time in the Philippines, it was disconcerting to see so many people go jobless. Companies and banks were closing down one after another. One of the banks closed on one of my accounts. It was nothing I experienced before, with all the suddenly difficult financial adjustments I had to do. I was not prepared for all these sudden disturbances. I went to see the manager of the bank and demanded, "Give me my money for my payroll." The bank just could not do it. I had to pawn some of my diamond jewelries to pay the salaries of my staff! There was no such thing as government bailout in a Third World country. It was a financially exhausting time for everyone. It must have been as difficult for a lot of other companies. Many companies closed. Unemployment skyrocketed! The same similar effect occurring from a society that has become turbulent. No one is really prepared for these sudden bursts of turbulence in a country that has gone really wrong. At that time, you just remedy the problems that come along and live with all the suffering. Before I learned TM, it was like working under constant storm in a turbulent country. After TM, the stormed just instantly calmed down and the sun started to shine. How refreshing!

The band members in Joey's party were from the group of 1,200 TM and Yogic flying participants that Maharishi had sent to the Philippines to quell the turbulence in the country. It was the square root of one percent of the country's population at that time in 1984. Trouble was really brewing in the Philippines. It was an opportunity to validate the Maharishi Effect. Maharishi saw all the signals of nature that was simultaneously occurring at that time, people protesting, typhoons prevailing and even a volcano was erupting at the same time. Maharishi then instructed all TM practitioners from around the world, "Go, as if the house is burning."

Maharishi proved then that the turbulence, including the violent typhoon and the climates changed! The volcano that was raging at that time subsided, after Maharishi sent a group of Transcendental Meditation and Yogic flying practitioners together in a downtown Manila hotel. Strange, the turbulence did subside with the

1,200 TM practitioners that Maharishi had sent to the Philippines in 1984. The transition to the newly formed government was as peaceful as many can recall. The TM group stayed until 1986. Not a single gunshot was fired from both sides, from the protesters and from the military that came to oversee the large, noisy rallies that went on in the streets of Manila. A new government was hurriedly formed in a peaceful, orderly manner.

President Reagan was the current President of the USA at that time in 1985, he sent a military helicopter which picked Marcos and his family and flew them to the island of Hawaii. Marcos left the country and the new government took control. The people uprising settled down. This was a validation of Maharishi's Unified Field Theory, the Maharishi Effect. The simple natural phenomenon of order that occur when the square root of one percent of the population practice the Maharishi's technology of the Unified Field in groups together, the turbulence in the country subsided, like boiling water that instantly cooled off and simmered down.

After that evening party with Joey, the Unified Field group started to call my office almost daily. I was too busy to answer and entertain their calls. I had many priorities. My tasks in the office seemed endless. It was very tough for a young single mother managing a company composed mostly of male employees. It was exhausting managing all the male engineers and construction workers who were mostly men. There were always the difficult male ego issues I had to cope up with. There were always employee issues I had to confront. To top it all, I had to come up with the money for the regular weekly payrolls and payables occurring daily. I had to come up with alternate funds as many banks were unable to release any cash at that time. It was a life that was just as tough as any one can imagine. I wondered how I managed as a young mother alone, who survived working during a very turbulent period of history. Hard work alone would not have paid off without my regular TM practice!

Due to limited time to accomplish unlimited work, it was impossible for me to entertain and consider other social calls outside of my work hours. I could not find time for any friendly encounter during those very busy times of my life. My life consisted of work during the day and caring for my children in the evening.

After learning TM, my busy office work reversed and turned around completely! I was alternately having fun at work or happy at home with my kids. The stark contrast made me start to wonder where all the good things come from the practice of TM, where did all the happiness and the joys come from after learning TM. Like the sun suddenly rose brightly and calmed down the constant storm.

I did not have a bit of knowledge about Maharishi before that, naturally, TM had limited importance to me then. By accident, on my way out of my office one afternoon, I overheard my secretary arguing with someone on the telephone. "Unifield? No, she is not in," looking up at me and shaking her head. I thought it was the one of our construction suppliers, the Unifield Corporation whose call I had been waiting for. I took the telephone from my secretary's hand and inquired, "Is Ted there?" Ted was the president of Unifield Corporation.

"No, I am Stephen from the Unified Field band. We met at Joey's party, remember?" Stephen seemed very nice. I just could not refuse him. He was the spokesman at Joey's party. He was persistent and determined to meet with me. I hesitantly accepted his invitation for an appointment for meditation lessons. I accepted the invitation cordially. It was an unusual kind gesture I gave these young Maharishi band members. At least that was my initial thought. Twenty minutes was not a bother, so I thought at that time.

I hesitantly learned Transcendental Meditation the next few days. A lady came to my place and taught me TM one-on-one. That was after Stephen and his team gave me the preliminary lectures on Maharishi's Unified Field that was required prior to learning it.

When Joey found out I had learned TM, he said, "When the student is ready, the master appears. You cannot run away, even if you try. It is part of your destiny," he said joking and laughing as I told him the story of the corporation with similar name, the Unifield company that provided some construction supplies my company needed at that time. Nature has its strange ways. From then on, nature and destiny took over my busy life, It shifted from being difficult and busy, to fun time at work. The sudden change amazed me with so much wonder! I started to meditate, twenty minutes upon waking and after work. I was thrilled that I felt refreshed immediately after each TM program. I appreciated meditating in the afternoon because after a busy day of work, I become so tired and stressed out. After meditating, you feel so much more relaxed and refreshed. I felt like I was ready for another round of work. It was exactly what I would do if I had unfinished business to complete, which I had most of the time. The effect of my TM practice was more apparent in the afternoon, when fatigue would begin to set in. After working long hours, my chronic headaches would come back. My regularly recurring headaches instantly banished after TM. What an instant relief! I was better able to complete my days work load. I did not waiver in my decision making, as I had in the past.

After a few days of meditation, my recurring headaches not only disappeared, they never returned. I was less prone to fatigue. I gained more energy. I was able to perform more work in shorter periods of time. I came home earlier and enjoyed more time with my young children. My children gave me so much joy. Just being with them and sharing in their company was truly a fun time for me, in contrast to rarely seeing them during the day. I would depart for work very early and arrive home sometimes late and they would already have retired for the night. I never wanted missing dinner time with them.

At that time, my children had a nanny named Inday who lived with us for many years. She took care of my children. Every evening, she reported to me like a home secretary and accountant, keeping track of my home and domestic finances like a professional. Inday also began meditation. She was like a second mother to my children's domestic concerns. She became a part of the family, a meditating family with my children having become regular meditators too. It did not matter that there was no father figure around. We all grew to be very happy together. The children never seemed concerned that they had no father in the house. They were just so joyful when I would come home early and enjoy dining with them every night and go shopping with them every weekend!

My children couldn't help but notice how my demeanor had changed. They had to adjust to my breaking out in a song when I'd arrived home. This was in marked contrast to arriving home with a headache and a long face being exhausted from the day's work. They would even buy me some song booklets so I could remember the lines of the songs I liked to sing. At times, we would all sing together as a group, my kids out singing me most of the times.

My happiness grew with each day that passed, in contrast to sometimes worrying myself to sleep before I learned TM. Maharishi's TM really worked! Research said it works for everyone. My difficult and seemingly solitary life changed dramatically. It was truly due to my regular TM practice. My life became so much happier with my family. TM was like a lamp brightly lighting my way thru the dark.

Maharishi said, "It is like you are wearing dark glasses, all the scenery you see seem dark. With clear glass, everything becomes clear." Life truly became very clear and with so much more joy filling your heart all the way! You see life from a much different and happier perspective. It is not like life is dark with suffering. Maharishi laughs at suffering, usually he holds a rose stem and joyfully swings it back and forth like swatting a fly or like he was swatting away suffering.

At the office, all my staff started to meditate. I was overwhelmed by the change in the environment of my team. It became more obvious each one was more supportive of each other, in contrast to the old office rivalries. The complaints and bickering about who was paid more with less responsibilities and who was paid less with more responsibilities subsided. I became more content and satisfied with every one working at such a high level of performance. There were no more need to demand completion of assignments or break up confrontations. Everyone performed their tasks with minimal supervision. Productivity increased. Creativity to solve work problems easily just came out from everyone.

The office environment dramatically changed. It had taken a clear step towards exceptional excellence never seen before from everyone, even from the lowly workers who were thinking as good as the engineers themselves. There had been previously nothing obvious like it. There always were problems I had to deal with. Absenteeism also improved. Now, people enjoyed coming to work. They looked happier than the gloomy faces I previously observed in the mornings when they come in for work, that was before they learned TM. Before the group practice of TM, a drunken fistfight caused a worker to fall in the elevator pit and died. Constant arguments common among the construction workers did not happened anymore. I thought to myself, "What could have caused this change?" The only significant change was this, all the workers and staff, 400 of them, had started to meditate. The Maharishi Effect changed the whole work scenario!

To my surprise, the male staff would salute me when I passed by. This is in contrast to having them argue back at me at times. I was truly flattered by what was truly happening. On many occasion, in the evening, I would have to complete necessary tasks not completed and left unfinished that needed attention for the following morning. It was urgent that schedules of work and materials for the next day be prepared and ready, otherwise, workers will be on standby and not briefed what to do next day and a lot of time and money will be lost.

Now after learning TM, my staff would finish and complete the work themselves without being told, even staying late if additional time was required to complete the tasks. I was so happy I can now go home earlier than usual without any concern about the work next day. Before my staff learned meditation, it was disheartening to see everyone looking at the clock waiting for it to strike five o'clock in the afternoon. Then, they would rush out of the door uncaring about the work left behind. Work issues just disappeared. It was very peculiar.

I couldn't believe the dramatic change that had come over my organization. I found out that the same effects were being observed in other companies that took on with the practice this Transcendental Meditation. I began to believe Maharishi's statement that happiness is man's birthright. I witnessed so much happiness from all of my staff and workers, including the cooks in the canteen!

At that time, one of the biggest companies in Japan, the big Sumitomo Co. implemented Transcendental Meditation in the corporate human resource department. They reported the same type of change in personnel relations I had experienced in my company. Other world-wide companies implemented TM and confirmed similar results from the group practice of TM. I just duplicated the trend of the program with my team.

Oprah Winfrey, the famous American TV personality had all her staff meditate, she said the same transformation too! Oprah Winfrey's staff now practice TM with her in the office twice a day. She says, "Nine o'clock in the morning and 4:30 in the afternoon, no matter what is going on, we stop and we meditate. And that way of being "still" with our selves-coming back to the centre and recognizing that something is more important than you. It's more important than the work you are doing, brings a kind of energy and an intensity of energy, an intention we never had before. And you can't imagine what has happened in the company. People who have migraines, don't. People are sleeping better. People have better relationships. People interact with other people better. It's been fantastic." [15]

It is a natural phenomenon happening in all groups practicing the TM program together. "The magic starts from within," Maharishi repeated. Truly, the stress that has accumulated in the body and nervous system melts away with the practice of TM.

When I first met Maharishi, I was not as attentive as I had to listening to him, with jet lag and all. All the others in the audience were intently listening to him, some took notes and even wrote down Maharishi's lectures. After just one month into TM, I appreciated the relaxation I gained daily and did not yet consider TM as something of such great importance as to hold a whole range of spectacular benefits

---

[15] *Tm.org web page*

including solving climate change. I did not expect to sit and listen for hours to someone I haven't heard of and known till just a month ago. I just followed everyone going in the hall and listen to Maharishi talking for hours until midnight.

The following day, again, there was another lecturing session. During Maharishi's speeches, I sat half awake, half sleeping under his gaze. I couldn't keep my eyes open still with jet lag. I wanted to leave, but I was sitting at the center right under Maharishi's gaze just a few rows away from where I sat. I felt uncomfortable with him looking straight at me. I tried to move to the left as soon as a seat on the left was vacated. But people continued to come from the left and asked me to move again to the right so others could be seated. I found myself once again at the center moving both ways as more people came and went from both sides. In the end, there was no place to move right or left. I just sat uncomfortably back again right at the center under Maharishi's gaze. I really wanted to leave. It was uncomfortable as my head was falling from my shoulders and my mind was at the brink of the sleeping state, it was truly embarrassing.

Maharishi could see me losing the battle to stay awake. It was an uncomfortable feeling sleepy with someone staring at you. I whispered to the woman beside me. I met her on the plane. She was a famous Filipino TV anchor woman and newspaper writer. "Let's leave, Julie," I said. "I am really very sleepy." "So am I," she said.

"Okay, let us move slowly behind the chairs, so Maharishi won't see us," Julie softly whispered her instructions in my ear. Very slowly we moved out from our seats, nearly crawling on the floor behind the chairs. We were just three rows away from the podium where Maharishi was sitting. I could see Maharishi's eyes rolling and following us as we reached the exit door. Then he suddenly stopped speaking. Instead, we heard his voice booming over the microphone. "Where are you two going? Come back here in front," as he pointed to the middle of the front row, right in front of him. There was a loud burst of laughter from the audience. The roaring laughter from the audience, about five hundred in attendance, completely shook me out from my sleepy state. Embarrassed, we stopped crawling and we came to sit in the very front seat near Maharishi. We obediently sat there until Maharishi's finished all his long discourses, right to the very end.

Later someone passed me by at the hallway and said, "Maharishi likes you." I replied, "I own a school, maybe, that is why

Photo taken on January 12, 1985, author is next left of Maharishi

he took notice of me." The practice of TM in groups together, was what Maharishi sought for and wanted, that is why he paid attention to me. "I have a school with four hundred students," I said.

On the last day of the conference, before leaving for Manila, we were called for a brief audience with Maharishi. There were about a dozen of us Filipino participants. A group picture was taken. See photo above. I sat next to Maharishi on his left covered with the white shawl someone had previously put over me. With him so close, I took the opportunity to quell my curiosity and observed him from head to toe.

He covered his toes as he noticed I was looking at them. Maharishi wore a monk - attire made of ivory white silk, with no socks or shoes, only wooden sandals. He always sits in a lotus position. When we were all seated quietly and comfortably settled before Maharishi, he asked, "Philippines, what are your plans?" Then he was very quiet. No one replied. There was an uncomfortable moment of silence. He was looking very sad at all of us. I was touched by his sadness, so I broke the silence and said, "Do not be sad, Maharishi, we will bring you the seven thousand."

The seven thousand was the square root of the population of the world then in 1983. He smiled faintly at me and said, "If your intention is pure, nature will support you." He must have thought I can create a model for the Maharishi Effect as I run a school then. Having been into meditation just a few weeks, I barely understood the concept of the seven thousand or the group needed to create the super radiance effect, the Maharishi Effect that was needed for the world then in 1985.

This time, the number of people needed for super radiance in the world has gone to more than nine thousand. As the population increases, the group required for super radiance proportionally increases. Maharishi equates the super radiance effect with total positivity in the world. "All darkness will disappear and only goodness will flourish," he said.

During the course of the conversation, Maharishi was emphasizing the group effect. It will produce positivity and all problems will melt away. "The Philippines is the land of the rising sun," Maharishi said. Japan is the land of the rising sun and the Philippines is the pearl of the Orient. Maharishi was not correctly informed. He continued, "Let coherence start from there radiating throughout the world," he said.

There was so much negativity and poverty in the Philippines. That was the reason he was so obviously sad seeing all of us. No one seemed to have a plan. I came without one either as it was the first time I heard about the group effect and the plan to establish it.

A lady approached me after we left the room. She said, "You promised Maharishi the seven thousand. Do you know what the seven thousand means? They have to be practicing TM- Sidhi and Yogic flying in one place together!"

She stopped in front of me with a funny stare and playfully shook her head. She then hurriedly left giggling as I stood stunned in the hallway wondering in awe about the 7,000 she was talking about. The seven thousand meant nothing to me then.

For many years, fifty years to be exact, Maharishi has been trying to establish the group on a permanent basis. That was why the lady had that funny stare and had laughed at me at the hallway. I wondered if I could ever really do it as I promised Maharishi. No one has done 7000 to date on a permanent basis. Many decades has passed until now and still the magnificent effort continues to this day - it is still being done now at the central part of India and in Fairfield, Iowa, USA. Then, I had not the faintest idea what was meant by Sidhis and the Yogic flying. I was aware of only the twenty-minute meditation periods, the individual practice and individual effect. Maharishi has repeatedly emphasized the group practice and that the effect is exponentially powerful. "Meditate in groups, do the program together. The power is in the group meditation. Establish schools. Establish big groups meditating together. Do super radiance together. Do group, do group," Maharishi constantly repeated.

The individual effect and benefits of Transcendental Meditation can be immediately experienced after every twenty-minute practice morning and afternoon. It was really good, I experienced it. I had no idea about the group effect then, being into TM barely a month. I found out later that the same is true of a large group practicing together. The effect is exponentially much larger and far way beyond any description. The Maharishi Effect is a spectacular mind boggling bliss experience that the wonder lasts a lifetime.

Indeed, for me, the Maharishi Effect is the greatest and most interesting endeavor to validate. If it is possible to eliminate all negativities in life in one single stroke, who would not want to consider doing it?

I started to consider doing it. I wanted to personally test it. It was like wanting to change the common scenes of poverty you see around you or the news you hear about crimes occurring daily. Many sick people are crowding in the hospitals. Children lay scattered in the overflowing hospital hallways. In a Third World, the scenes are truly pathetic! Now, countries are having problems. Not only crimes happening. Terrorists are rising. Wars are occurring. Conflicts and fighting in many countries like Syria and Ukraine deeply troubling the West. In war torn Middle East nations, many families bear the bitter consequences of war. Refugee camps are filled with mothers and children. Multitudes of migrants flee war torn Iraq, Syria and terror ridden parts of Africa.

Storms are flooding communities. Severe fires soaring in many states. Suffering is too much to bear. When the human mind starts to be turbulent with people in groups protesting, governments begin to weaken and many times topple down. Many people die.

How lucky the next generations will be. We have the Maharishi Effect to end wars, to balance climates and to create a truly happy world. All that is needed is to form TM and Yogic flying in large groups together. It is an opportunity for governments and men of fortunes not to miss it. They have the capability to establish large groups and provide humanity the great benefits of the Maharishi Effect.

6

THE GREATEST EXPERIMENT DONE ON EARTH

In 1983, in Fairfield, Iowa, USA, several scientists worked with Maharishi and conducted a large group experiment on the effects of the 7,000, which was then the square root of 1% of the population of the world. The event was described as the Global Taste of Utopia which has become the greatest experiment done in human history.

That was a year before I began my Transcendental Meditation program. After reading the documented papers of the Global Taste of Utopia event, I was driven to validate the group effect in a small scale and see the real fact. I did not want to believe it at its face value, whatever was said about it. Many of my friends doubted that climate change or world peace can be solved by Yogic flying, something they have never heard about. When they found out I learned TM and was flying, my colleagues ridiculed me, "What is a structural engineer like you doing with this thing? The professional group I belonged to then had many highly learned engineers and some pursued PhD degrees in engineering from prestigious universities in the USA, they were not aware of something as spectacularly significant discovery like the Unified Field or the Maharishi Effect. Young as I was then, I started to wonder why the spectacular effect has not come to the attention of my Filipino colleagues in the profession. Usually, discoveries in the world that made exceptionally significant effects spread like wild fire. The experience of the effects of Transcendental Meditation and the Yogic flying was overwhelming, instead of wavering, I became more curious and driven to find out where all the good things come from the practice of Maharishi's programs.

Maharishi said the Unified Field is a science. Later, from research, I found it to spring from the unseen, unmanifest, non-material, root form of all material creation, contained in the science of particles I further discussed in Chapter 7, the science behind the natural character and behavior of the finest elementary particles of nature. It explained the approach to solve global warming and further clarify the general understanding of climate change. Besides climate change, it contains the model to transform the world from the prevailing negativities occurring around the world to positivity in one single, easy process just by gathering TM and Yogic flying participants in groups together. Yogic flying occurs at the point of maximum brain wave coherence, then the body lifts up defying gravity. With the discovery of the Unified Field, it was found that the human mind can now handle gravity instead of just staying in transit.

"It is too good to be true," some of my friends say. At that point in time of my life, I was never a seeker or a follower of any group or basically a joiner of anything. Being a single mother with three small kids to support, I was just so busy working hard and providing for them. However, the documented result of the Global Taste of Utopia appeared truly important to miss out on it. I felt that it has to be proven true. How could you miss to validate it considering the significant contributions it can give to everyone, nothing like it was ever found before! Maybe I could establish a small school and observe the results. I proceeded to establish few groups to determine in my own mind whether the effects are the same when repeatedly put to practice.

I found out that around the school environment, the effect was very much the same from group to group. The students in the schools surprisingly increased their IQ so easily in just a short period. Their behavior significantly improved and they became happier. The climates in the surroundings improved significantly. Economies rose. Peace prevailed. It showed similar results, the same positivity illustrated by the global event Maharishi organized in Fairfield, Iowa, USA. I was overwhelmed by the excerpts of Maharishi's Global Taste of Utopia experiment in 1983, I could not believe a transition like this could possibly occur. The change from negativity to positivity was documented in the following pages.

Many will doubt the same claimed effect unless they do it themselves too, like I did in a small scale, about 500 students each time. It was really an awesome scientific discovery that Maharishi was able to connect the theoretical particle science known to physicists, and relate it to the attributes of the Unified Field found enlivened within the silent level of the human mind in large groups together. In my opinion, it was a quantum leap discovery of knowledge in the history of mankind.

Some scientists stipulated that human consciousness affect global events. Einstein, in some of his papers, it said that the mind of man determines the environment and Maharishi proved it to be so, as illustrated in the global event in Fairfield in 1983 and the small events that I had created for my reference.

Einstein's postulate was truly significant. It foretold Maharishi's confirmation of the Unified Field effect in the environment. It was Maharishi, a physicist himself like Einstein, who confirmed that this theory of the Unified Field occurs within the mind of man at the level of silence. Einstein postulated the energy field in the vacuum state and called it the quantum state in physics. Einstein also referred to it as the Unified Field. Quantum physics revolutionized the energy fields of nature at the finest forms of the atoms. Maharishi postulated that Unified Field contains the finest of all particles of nature, finer than protons, photons or photinos, electrons, neutrons and neutrinos and beyond finer particles. Maharishi referred the finest particle of nature to be the "sap" of the universe. Like the sap inside a tree, it makes the tree grow, blossom and bear fruits. The finest of all particles in nature, the "sap" that Maharishi referred to, I called it the anima solo particle, the AS_U, for my reference. It is the particle that nourishes and animates the whole universe that contains the blue print of the total knowledge of natural law that nourishes the whole of humanity when established in large groups together. It is enlivened at the level of silence within the human mind during the practice of Maharishi's Transcendental Meditation.

Both Maharishi and Einstein were physicists who tried to transcend the untapped power behind the fundamental and primal particles of nature. Maharishi found it to permeate all creation and found also to linger within the human mind.

Maharishi found the elusive quest for unity in nature where Einstein failed. Maharishi had with him the ancient knowledge of meditation from the east and it was the tool Maharishi used to validate the theory behind the concept of what was behind unity and singularity in nature. Maharishi had the knowledge of the Vedic literatures from the ancient Vedic civilization of India handed down to him from the Himalayas by his master, Guru Dev whom he served for nearly twelve years until Guru Dev passed way in 1953.

Einstein, coming from the west, had no knowledge of meditation, in contrast to Maharishi who came from the Himalayas known to be the seat of the eastern and the ancient Vedic tradition that referred to the secret knowledge of the functioning of the mind during meditation.

Einstein pursued the Unified Field[16] for many years but was not able to determine that the Unified Field could co-exist with gravity. Gravity, the weakest force in nature posed so much resistance and in all of Einstein's experiments, he found gravity constantly stayed in transit. Without a background in meditation, Einstein was not aware that the laboratory of unification of the fundamental forces

---

[16] *Einstein: Life and Times, 1971, Ronald W. Clark*

of nature could occur within the human mind. Maharishi located the Unified Field within the human mind because he was familiar with the mental procedure of Transcendental Meditation that he learned from Guru Dev, his teacher.

Maharishi studied for more than a decade under Guru Dev in the deep forest of the Himalayas. From Guru Dev, Maharishi learned the much sought after secret knowledge of the functioning of the human mind. "At the vacuum state," as Einstein called the quantum state, "there appears to be nothing." It is unseen and unmanifest. Like when solid ice turns liquid, then heat the liquid water, it turns to gas. The water molecules turned gas in the air becoming invisible, unseen, unmanifest and non-material but the hydrogen and oxygen component of water in the air remains the same. Einstein developed the idea of the breaking the atoms and his formula was the fore runner in the development of the atomic bomb. The hydrogen and oxygen component of the water becomes the unseen air we breathe, and that which gives life, only the visible form of the solid ice crystal changes. It turns to gas we call air. The finest elementary particles of nature are unseen, unmanifest, but none the less they exist. The alchemy is produced by changes in temperatures.

The sun produces heat in our galaxy proportional to the distances of the planets moving around it. The interplay of atoms around each planet are unseen and unmanifest. The finest particle of the nature contains the blueprint of the total potential of natural law moving around everywhere in the universe. Hydrogen and oxygen breaks down to smaller particles to the finest, smallest and purest particle, the solo particle. I referred to it as the solo particle because Maharishi said, "it creates from itself, by itself and within itself." It does not need another particle to co-exist with it, like protons and neutrons to produce electricity or two parts hydrogen and one part oxygen to produce the water particle or the same air particle. The solo particles at the level of the Unified Field can be drawn within the human mind, at its most silent state. The quantum field, the Unified Field of all the laws of nature that Einstein referred to, Maharishi found it to linger within the human mind at the level of silence. This nothingness, this finest particle form is the seed form of all the laws of nature. It is enlivened at zero excitation of the human mind. It is like the theory of everything, espoused by Stephen Hawking, a contemporary physicist who referred to the Big Bang as the origin of the universe. Hawking stipulated that the universe sprung from a vacuum state of singularity, a single form of concentrated unseen energy, from nothing manifest, from zero to diverse quantum materialization of the universe. Hawking said it did not need a God to create the universe. The laws of physics did it. The laws of nature created the universe.

Maharishi stipulated that the vacuum state, at zero excitation of the mind, there exists the finest energy of nature from where all of the energies of the universe spring from. It is the pure field of positivity, it contains the natural nourishing behavior of natural law, it cannot create any negativity. It unifies all of creation in one singular finest particle that creates an invincible field, the Unified Field. This knowledge elevates man to the highest level of nature, to the most supreme seed form of natural law. Man is the ultimate and supreme conductor of nature at zero excitation within the human mind. Man has become the conductor of the Unified Field, the channel and conduit of natural law.

What Maharishi has said is that our material body and mind has the Unified Field, the finest form of energy field of nature within man that is governing and administering all the functioning of nature. It is the energy that gives harmony to every moving planet, and the same energy that gives life to every cell within the human body.

In a car, the energy starts as the ignition is turned on, switch on the car key, the car starts to move. In man, as soon as the mind quiets down, the Unified Field starts to get enlivened, the human mind begins to tap the main switchboard of natural law. Maharishi's Transcendental Meditation quiets the mind in the most easy and natural manner.

The Unified Field begins to be enlivened within man during meditation and reach the deepest level of silence. It is like many rivers going back to the ocean. It is like a field of energy, all the energies of nature go back to the ocean of energy. The Unified Field within man is the same energy outside functioning in the vast silent universe. Like a tree, all the branches of energy unite at the root.

The human mind at the waking state cannot easily quiet down. Without meditation, it is impossible to quiet the mind. There are always thoughts. Once the eyes closes there are always many thoughts that come. The silent state of the mind, soon reached, creates the most blissful state and becomes the highest level of life in enlightenment man can attain in his lifetime.

It truly is beyond common comprehension and describes it as the state that is far beyond anything. The vacuum state in physics is a state where no material particles seem to exist. Nonetheless, it contains the finest unseen particle of nature and becomes the most powerful energy particle when contained in large groups together.

They do exist within the human mind at the level of silence and their effect is illustrated in the Maharishi Effect, in the Unified Field effect that Maharishi has foretold.

It is an energy field, like electrical field with unseen particles. Beyond electricity, magnetism, and gravity, the junction point of all energy united at the source, at the Unified Field energy that exists also within the human mind.

Maharishi said it is the most supreme state. After validating the effects on a small scale, using the small schools that I have established, I experienced similar results like those evidenced by Maharishi's large - scale experiment in 1983. The results were the same, in large or small scale project. The results of the large - scale experiment are attached here in the following pages described in the Global Taste of Utopia experiment in Fairfield Iowa, USA following herein.

From December 17 through January 6, 1983, more than 7000 experts in the Maharishi Technology of the Unified Field from 50 countries, assembled in the presence of Maharishi Mahesh Yogi at the Maharishi International University in Fairfield, Iowa, USA to give the world a taste of utopia.

Research scientists at the Maharishi University of Management predicted in early December 1983 that the reality of the Global Taste of Utopia would be proclaimed by improved relations between nations, greater cooperation between political parties, increased social harmony, signs of worldwide economic recovery, and reflect growing optimism, confidence and self–reliance, along with a general sense of well being in the family of nations.

Above all, it would be the head of state, the embodiment of national life, who would be the first to enjoy the sweet taste of utopia rising in his country and the world. Asked about the anticipated results, Maharishi said, "Anything that is good will rise; anything that is not good will vanish, as darkness disappears with the first ray of light."

World events during the three week assembly fulfilled the expectations of Maharishi and the scientists, clearly demonstrating the collective performance of the Maharishi Technology of the Unified Field in one place on earth, by one group of 7000 individuals. This number is approximately the square root of one percent of the world's population, at that time in 1983.

The number is sufficient to enliven the Unified Field of all the laws of nature in the whole world consciousness, and enrich all trends of life everywhere with the beautiful and nourishing qualities of the Unified Field.

Before the assembly, Maharishi predicted that rain would fall in Brazil, and indeed the long awaited rain did come in Brazil during the assembly of the 7,000..

The following are the documented results of the large scale Global Taste of Utopia event.

World Press Reports Documenting the Reality of the Global Taste of Utopia

"...let's be grateful to Maharishi Mahesh Yogi, the 7000 students of MIU and the several thousands of participants who flew in to the Fairfield, Iowa campus for a three week "Taste of Utopia session...." - The Washington Post, Washington D.C. January 10, 1984. "....we are impressed with what a special year it has been, what a change from last year end."

In particular, government demonstrated that it can do some things right, and was rewarded with improved public confidence restored economic growth, reduced social tensions and improvement in the U.S political influence in the world." – Wall Street Journal, New York, December 30, 1983.

"The world's stock markets were buoyant and governments are once again talking about economic growth...after four years of gloomy prognostications, there is, at last something to cheer about" – The Sunday times, London, January 8, 1984.

Increased Harmony in International Relations

"The New Year is coming in the signals from both East and West suggesting that the chill...between the two power bloc maybe moderating" –Financial Times, London, January 4, 1984. More than 60 nations announced initiatives to improve relations and strengthen ties with other nations as a wave of goodwill and friendliness swept the globe during the three weeks of the Assembly.

This was particularly evident in the Middle East, Central America, Southern Africa and Central Europe - regions of the world where greater harmony and stability are most desired by the family of nations. With the increased progress towards December 17 to January 6 as compared with the previous three weeks, the New Year began on a rising tide of hope for a more stable and peaceful world.

U.S. and Soviet leaders announced their desire to foster friendly relations. Relations between East Germany and West Germany..."seems to be brighter for many years..." after

"...extraordinary momentum in strengthening the links...."–The Times, London, December 29, 1983. The South African prime minister saw "prospects of better relations with neighboring African states," after marked improvement in relations with Angola and talks with Mozambique to establish a policy of peaceful co-existence.

Argentina, Venezuela, Columbia, Mexico, Panama, El Salvador, Honduras, Costa Rica and Guatemala worked to establish peace in Central America.

Improved relations were also seen between the following nations: Israel and Egypt, Syria and the USA, Saudi Arabia, Syria and Lebanon; Iran and the USA; Iraq and Arab Emirates, and Egypt; Egypt and Iraq; Cyprus and Turkey; Morocco, Mauritania, Algeria, Tunisia and Libya; Great Britain and Argentina; Guatemala and the USA; Nicaragua and the USA; Argentina and Chile;

The European Economic Community and the Andean Pact Countries; USA and Mexico; Greece and Romania; Yugoslavia and France; Holland and China; USA and Yugoslavia; Spain and France.

The Soviet Union and Malta; South Korea and the Soviet Union; China and USA; Canada and China; Malta and Libya; East Germany and China; Turkey and South Korea; Japan and the Soviet Union; USA and Japan. Australia and the Association of South East Asian Nations; ASEAN and Brunei; Japan and China; USA and Laos; Ivory Coast and Ghana; Kenya and Tanzania.

During the assembly of the 7000, many world leaders appeals for peace in the family of nations, including the leaders of the USA, the Soviet Union, China, Great Britain, Lebanon, Rumania, Greece, East Germany, Poland, Libya and Malta and also Pope John Paul II.

> The Pope said, "...there is one good that humanity together hope to taste, that there is peace. Peace...requires from man the sincere force and generous inner renewal." January 1, 1984.

Greater National Strength and Unity

A new feeling of confidence and optimism on a national scale was reported in many nations, including the Soviet Union and the USA with US optimism at its highest level in ten years. "...Americans approach 1984 optimistically feeling that the near future will be significantly better." - The New York Times, New York, January 1, 1984. "The fact is Ronald Reagan has made the country feel so good about itself." – The Washington Post, December 29, 1983.

7000 participants of the TM-Sidhi, including the Yogic flying program in the presence of Maharishi Mahesh Yogi, at the Maharishi University of Management, in Fairfield, Iowa, USA, from December17, 1983 to January 6, 1984 that created the global TASTE OF UTOPIA event validating the Maharishi Effect.

Increased popular support for national leaders and their policies was seen in the USA, Great Britain, India and Turkey. "It's been a Reagan Year." – The Washington Post, December 29, 1983. "The President is riding the tide of good feeling." – Time, December 26, 1983. Time upheld US President Ronald Reagan and Soviet Premier Andropov together as "Men of the Year."

The BBC selected British Prime Minister Margaret Thatcher as the "Woman of the Year." Significant progress was made towards reconciliation of Israeli and Lebanon, with calm prevailing at the New Year. A formal agreement was reached between the Lebanese government and the rival groups, and endorsed by Syria withdrawing troops from Lebanon. Israel, on a comprehensive security plan agreed to halt fighting in Beirut and the surrounding mountains. "A new spirit of unity and cooperation emerged among different interest groups and political parties in Ireland, Malta, Turkey, Israel, Chad, Sri Lanka, Japan, Argentina and Canada. Local press in Cyprus reported a "peace wave."

Increased Freedom and Tolerance

New steps were taken to end corruption and to strengthen and streamline government in the Soviet Union, Turkey, Bulgaria, Zambia, China, and Japan. Elections were held or announced in Bangladesh, Pakistan and Lesotho. Nationwide strikes were settled or averted in Great Britain, the USA and Bangladesh.

Major educational reforms were initiated or announced in Great Britain, Soviet Union and the USA. Freedom was granted or promised to prisoners held by the state of Mozambique, Tunisia, Zimbabwe, Angola, South Korea, the Philippines and Poland. New steps were taken towards recognizing and restoring the rights of minorities in Hungary, Egypt, Israel, Iraq, Uganda and Sri Lanka.

Indications of Worldwide Economic Recovery

"The chill has gone from the economic air…Whereas, in summer, even perennial optimists were infected by fashionable doubt. Few now question whether recovery from recession is either genuine or lasting." – Times, December 29, 1983.

Stock markets throughout the western world unanimously agreed…the recession is over, recovery is continuing, company profits are blossoming, inflation is wilting…records fell like nine points yesterday." – The Daily Telegraph, London, January 6, 1984.

Everything suddenly seemed to come together....It was as though investors everywhere had made a New Years resolution that the recession was over." – Times, January 6, 1984.

"The world stock market boomed during the Assembly, reflecting the confidence and optimism felt on a global scale.

Prior to December 17, 1983 (the day when the Assembly began), five of the nine major markets were declining. Within two days of the assembly's start, all the nine markets were rising sharply.

MIU scientists found that the increase in the number of people participating in the Maharishi Technology of the Unified Field at MIU during the Assembly had a significant influence on the trend of the world stock index, as indicated by the Time magazine series intervention analysis.

- Previous records were broken and new records were set repeatedly in the stock market indices in Japan, (seven days in a row).
- West Germany (breaking a 23- year record), Great Britain, France, Holland, Switzerland, Australia and Singapore.
- Stock markets boomed in Hong Kong, Mexico, Canada and the USA with record trading volume in New York.
- Growing economic strength was seen in many nations as evidenced by falling unemployment in Great Britain and the USA, best labor market recovery in 35 years.
- Dramatic rise in the composite index in the business activities in the USA and Great Britain.
- Exceptional boom in retail sales during the holiday season in the USA and Great Britain
- Achievements of economic targets ahead of schedule reported in Brazil, China and the Soviet Union.
- Lifting of economic sanctions and barriers against other countries by Nepal, South Korea, Japan, Turkey, Jordan and many European nations.
- Food grain production in India reported to have reached a record high
- Signing by E, E. C. and Andean Pact Countries, "one of the most important trade agreements, after years of financial crises" – BBC World Service, December 18, 1983.
- Lowest interest rates on loans to developing countries from World Bank
- Large loan awards from International banks to Zaire, Uganda, Madagascar and Mexico.

- Revised global economic forecasts by both the International Monitory Fund and the Organization for Economic Cooperation and Development, indicating a better economic future than had previously expected. The O.E.C.D. reported that almost all of its 24 member states were now enjoying renewed economic growth.

Expressions of Heads of States Indicating Rising Strength and Coherence in National and World Consciousness

- December 19 – "I have always believed that your army is really doing its job if it never has to fight." – President Ronald Regan
- December 20 – "Confidence is in the air and with good reason." – US President Ronald Regan
- December 22 – "I feel fine. I don't think I ever felt better."- US President Ronald Regan
- December 24- "…trusting in God is the one sure answer to all the problems we face."– US President Ronald Regan
- December 25 – "We in the commonwealth are fortunate enough to belong to a worldwide comradeship. Let us resolve to communicate as friends, in tolerance and understanding. Only then will we make the message of the angels come true. Peace on earth, goodwill to men."– Her Majesty, Queen Elizabeth II
- December 25 – "...a new world war can be prevented so long as all peace loving people unite to work for peace."- Chinese Premier Zhao Ziyang
- December 28 – "I accept responsibility for the bad as well as the good."- US President Ronald Regan
- December 30 – "This is only the beginning of the revival of Britain…We want men and women to set their hopes high and to carry them to reality." – British Prime Minister Margaret Thatcher
- January 2 – "I believe in personal rapport and mutual trust with leaders of other countries." Japanese Prime Minister Nakasone.
- January 6 – "…all of us can look forward to 1984 with even more confidence." US President Ronald Regan.

Decreased Negative Trends, Increased Positive Trends

Violation of the laws of nature is the basic cause of all problems and suffering. The decrease in negative trends observed in many parts of the world during the Assembly at MIU has demonstrated that the stress accumulated in the world consciousness due to the violation of the laws of nature by the whole worlds population can be neutralized through the collective performance of the Maharishi Technology of the Unified Field.

- Traffic fatalities sharply reduced during the Christmas season in the USA; The New Year's weekend was reported to be the safest on US highways in 35 years
- Traffic arrests and accidents during the holiday season in Great Britain decreased
- No fatalities occurred when an earthquake shook the entire Japanese Archipelago
- Terrorist plots were foiled in Hong Kong
- All aboard survived a jet that over shoot the run way in Malaysia
- All survived a fire that destroyed a Nile cruiser in Egypt.
- Dramatic drop in crime rate was reported in China.
- Radioactive fallout over Britain was reported to be the lowest since the 1950's
- International drug smuggling rings were broken by authorities in USA and Spain
- Forty percent drop in incidence in terrorism was reported in the USA
- Upsurge in traditional religious values among youths was reported in the USA.

Increased Balance in Nature: Stabilizing of World Weather Patterns

- Heat wave in Europe and cold wave in the USA were seen by meteorologists to have restored balance in world weather.
- El Nino, the warming phenomenon that disturbed global weather for more than a year was reported to have disappeared that year.
- Maharishi promised the people of Brazil that rain would come when the number enlivening the Unified Field reached 7,000 at MIU, the long-awaited rain did come.

Major Discoveries of Natural Resources

- Diamonds discovered in Bhopal, India.
- High-grade uranium discovered in Canada
- Large coal deposits discovered in India.

Important Scientific Discoveries

Maharishi has predicted that with the influence of intense coherence generated during the Assembly of 7000 in December 1983, deeper insights into knowledge would be gained. This happened on both level of approach, objective and subjective.

The charts that follow are based on the Scientific Research Documenting the Global Taste of Utopia from December 17, 1984 to January 6, 1984 at the Maharishi University of Management in the presence of Maharishi. Another 7000 event was conducted in India in 1996 and indicated similar trend from negative to positive trends. There were two events of 4000, one in Europe that ended the first Iraq war in 1990 in just a few months in contrast to the Iraq war after 911 where the war lasted more than a decade. Another 4000 event was organized in Washington D.C. that reduced crime rate down to 18%. In all the organized events positivity prevailed as in all other smaller groups done.

There were several 2,500 group events organized at the Maharishi University of Management in Fairfield, Iowa, USA at the Golden Dome of Pure Knowledge where all the participants were accommodated. Having experienced the Maharishi Effect many times, each time similar positive trends occur, the effects are conclusive to be true and correct.

The results from the similar smaller groups that I had personally organized in smaller groups of 400 truly confirmed the positive results. The feeling of bliss within is so overwhelming that the same feeling haunt me constantly and wish the group effect can be made on a permanent basis.

I have now grown old, but if I can go back to being young and strong again I will do the same thing over and over again. This time, if I had the means, I will put all effort to make it a permanent event. Leaders, responsible groups and particle scientists should consider looking more deeply into the Maharishi Effect and witness the nourishing aspect of natural law in the daily life of everyone.

Climates will balance. Storms, tornadoes and hurricanes will fade away. Economies will rise. Wars will end. Positivity will prevail and all negativities will become encounters of the past that will never be allowed to happen again. Particle science unfolds the behavior and character of the finest elementary particles of nature occurring within the human mind.

In the eighteenth century, only three particles were considered fundamental inside the nucleus of an atom - electrons, protons and neutrons. To date, hundreds of particles have been discovered. From electrons, protons, neutrons and many other particles were discovered like neutrinos, photons, positrons, fermions, bosons, muons, pions, quarks, W and Z particles, gravitons and the like. These particles create a field like the electric field with protons and neutron particles, gravitons create the gravitational field or the magneto particles create the magnetic field. The AS_U particles create the Unified Field enlivened at the silent level of the human mind.

Different particles were uncovered from the interactions of the four forces electromagnetism, strong force, weak force and gravity. The importance of these discoveries are the benefits that mankind obtain. Nuclear energy can be used as sources of electric power with the exception of creating the atomic bomb that creates destruction. That is why the development of nuclear energy is controlled by nations in the prevention of it being used for destruction.

Scientific theories are proven correct when the observed results show the same results repeated each time. More than fifty times, Maharishi has called for large group events around the world and fifty times it demonstrated the same positive indisputable results. Countless small group events also proved the results have the same positive effects in small scale.

No wonder after fifty similar experiments validating the Maharishi Effect, Maharishi described heaven on earth and endeavored to prepare his master plan to create it described in Chapter 12. The Maharishi Effect is an amazing and significant discovery of the 21st century referring to the transformation of man and his environment. It needs to be known by all. No discovery to date brings forth a positive change that effectively benefits the individual, community, country, the whole world and the planet.

The Maharishi Effect is an effective strategy for global transformation. I tested the same theory several times and found the same results. That is why I can now confirm and prove Maharishi's theory correct. Anyone can do the same, form a group and make them practice Transcendental meditation together. They will be surprised to find the results bring the same positive effects.

The global Taste of Utopia large experiment conducted at the Maharishi University of Management in Fairfield, Iowa, USA in 1983 showed the ultimate benefits mankind can benefit from. Attached Chart 1 to Chart 10 illustrates the documented results of the Global Taste of Utopia in graphic form re-created from the original format.

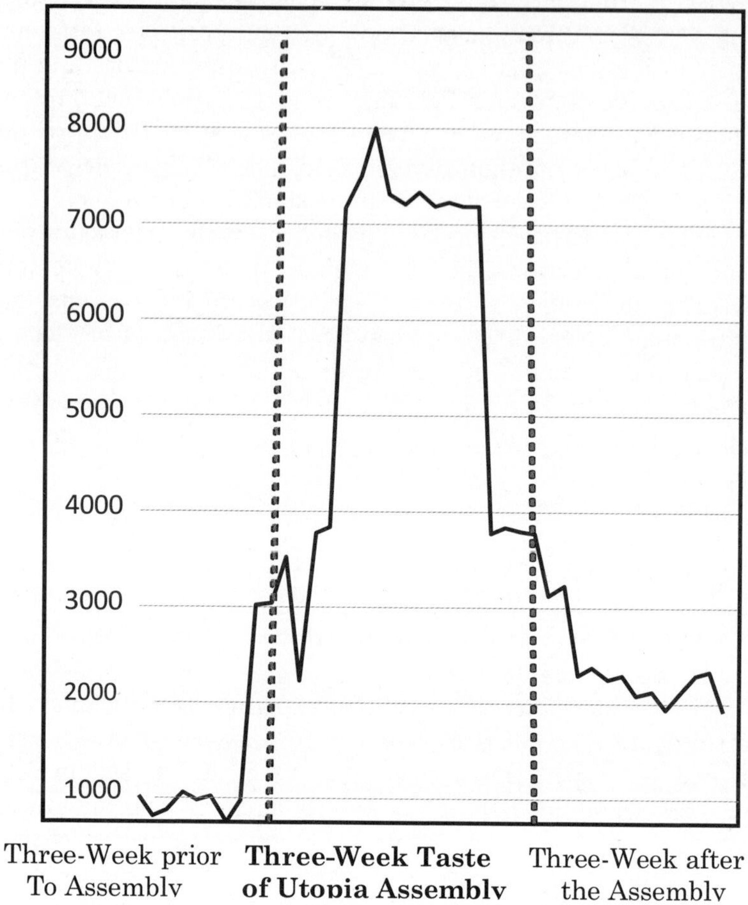

Chart 1 – *Number of Experts in the Maharishi Technology of the Unified Field*

Beginning December 17, the size of the coherence creating group at MIU increased to over nearly 7000, the square root of one percent of the world's population.

On January 6, the Taste of Utopia Assembly ended and the number of experts in the collective performance of the Maharishi Technology of the Unified Field fell far below the number needed for positivity in world consciousness.

Source: Capital of the Age of Enlightenment

Percent of Total Events as Rated for Degree of Conflict

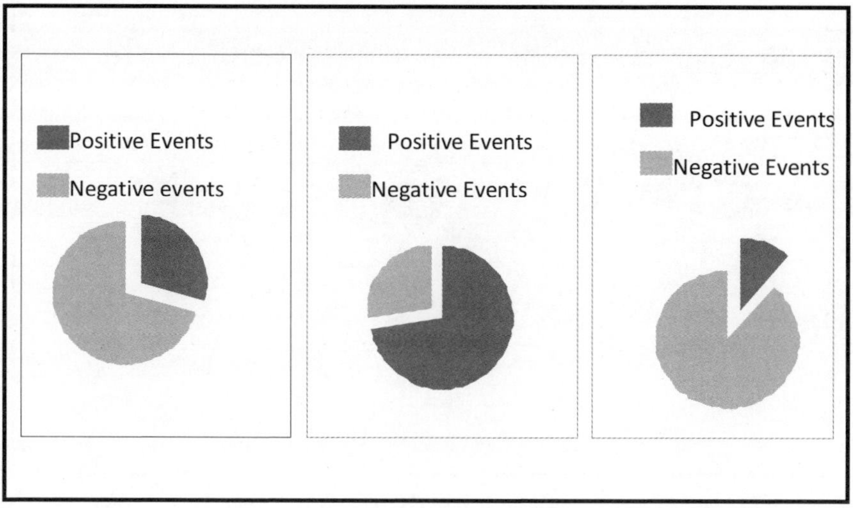

Three-Week Period Prior to Assembly | **THREE-WEEK TASTE of UTOPIA ASSEMBLY** | Three Week Period after Assembly

Chart 2 – *Increased Vitality and Positivity of the Heads of States*

According to Maharishi's absolute theory of government, the government is the innocent mirror of the nation and the head of state reflects by his speech and actions the quality of national consciousness as exhibited by more positive, evolutionary statements and actions of heads of states throughout the world, and by more national and international support for their policies and leadership. *After the Assembly, the quality of statements and actions of the heads of states and of the support they received reverted towards less positivity.*

Source: Content Analysis of *New York Times*

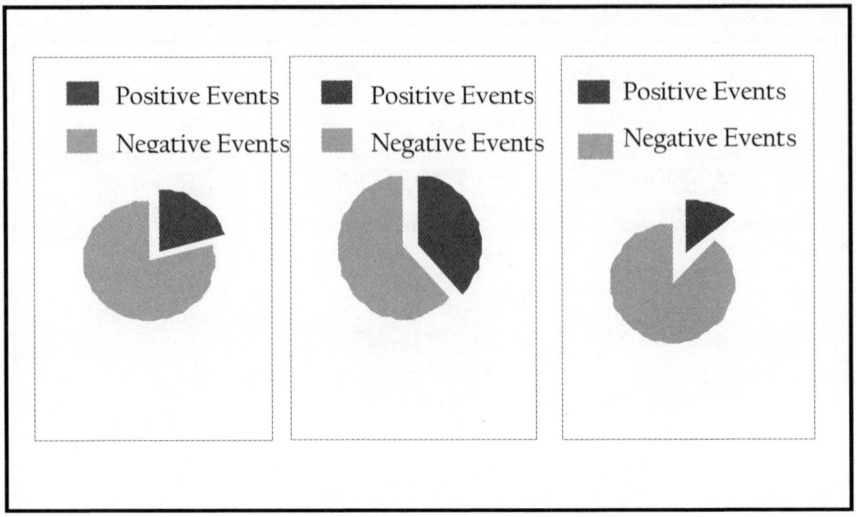

Three-Week Period Prior to Assembly | **THREE-WEEK TASTE of UTOPIA ASSEMBLY** | Three Week Period after Assembly

Chart 3 – *Increased Positivity of Events in Situations of international Conflicts*

During the three-week period of the Taste of Utopia Assembly the balance of negativity to positivity in events pertaining to international conflicts in the double-spot areas of the world shifted significantly towards increased positivity. *After the Assembly the balance of events reverted towards increased negativity.*

Source: Content Analysis of *New York Times*

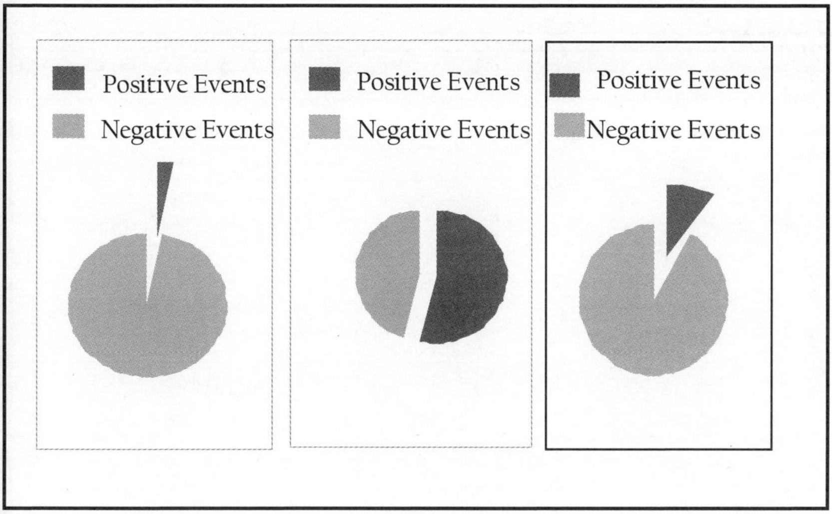

Three-Week period prior to Assembly	**Three-Week Taste of Utopia Assembly**	Three-Week period after Assembly

Chart 4 – *Increased Positivity of Events in the Lebanese Conflict*

During the three-week period of Taste of Utopia Assembly, positivity of events in Lebanon substantially increased, seen as progress towards peaceful resolution of the conflict. After the Assembly, the situation quickly deteriorated.

(Note: The content analysis was conducted by a team comprising members of various interest groups in Lebanon)

Source: Content Analysis of An-Nahar
(Major Lebanese Newspaper)

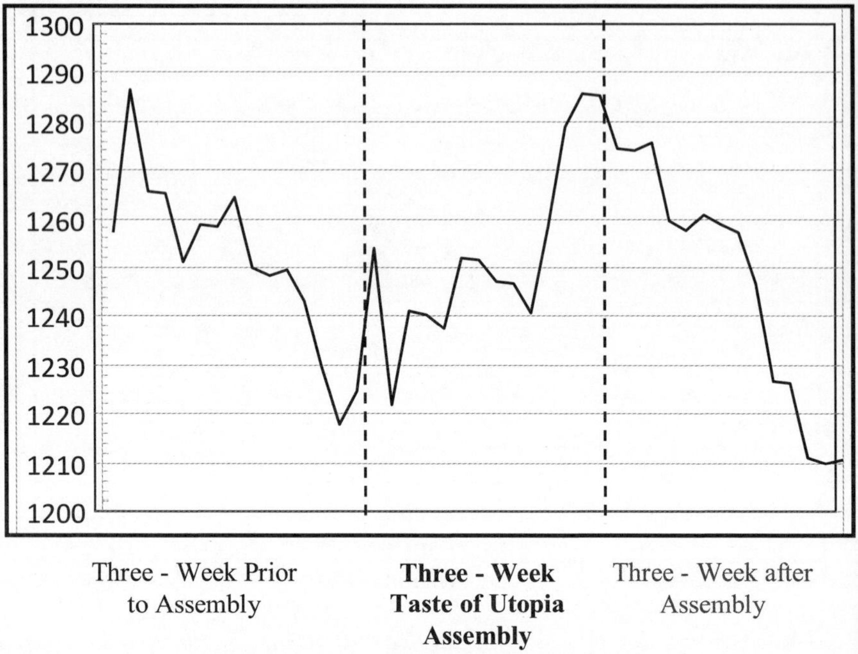

Chart 5 – *Stock Market USA*

Dow Jones Industrial Average

The Dow Jones Industrial Average showed a marked increase during the Taste of Utopia Assembly, reversing a prior, steep downward trend.
After the Assembly, the Dow Jones immediately reverted to a downward slide.

Source: Wall Street Journal

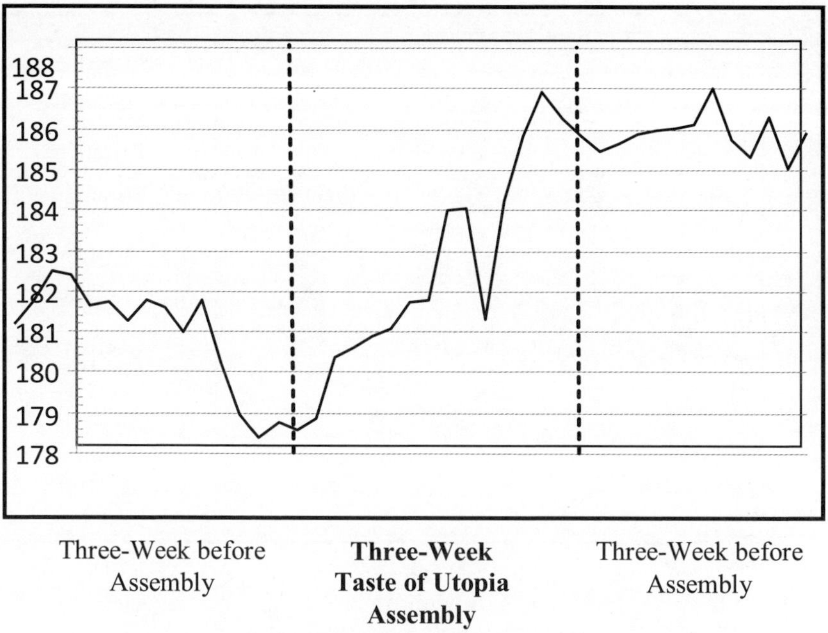

Chart 6 – *Rising World Stock Index*

World Index of Capital International S.A, Geneva

During the Taste of Utopia Assembly rising coherence from MIU created a wave of confidence and optimism throughout the world, seen as the rise in World Stock Index, a composite of 19 countries. *After the Assembly the Index resumed a downward trend.*

Source: Wall Street Journal

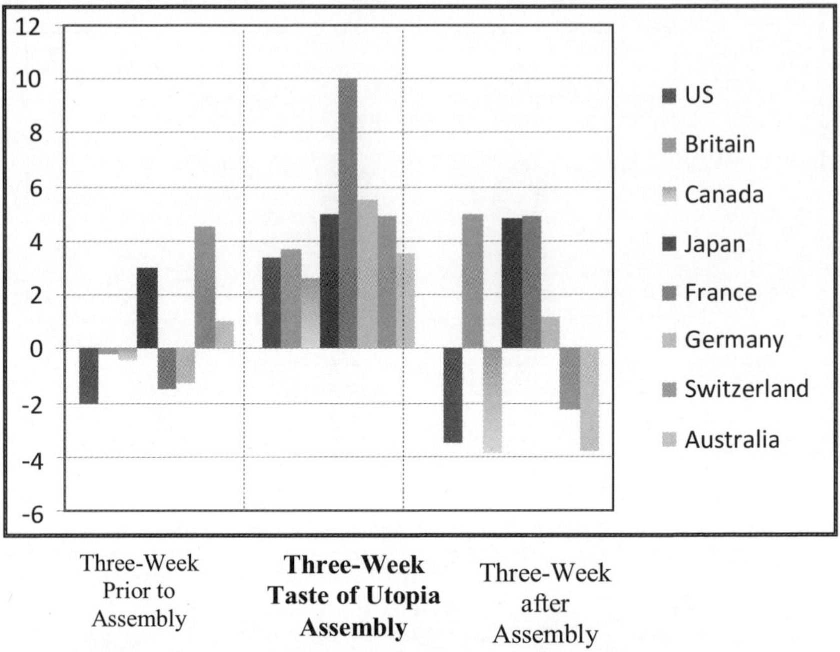

Chart 7 – *Simultaneous Increase of Major Stock Indices*

Percent Change over Three-Week-Period

During the Taste of Utopia Assembly, the major stock markets of the world increased simultaneously, indicating balanced economic growth worldwide. *After the Assembly, the same major stock markets reverted to a pattern similar to that seen prior to the Assembly, with some increasing and some decreasing.*

Source: Capital International S.A Geneva
(Wall Street Journal)

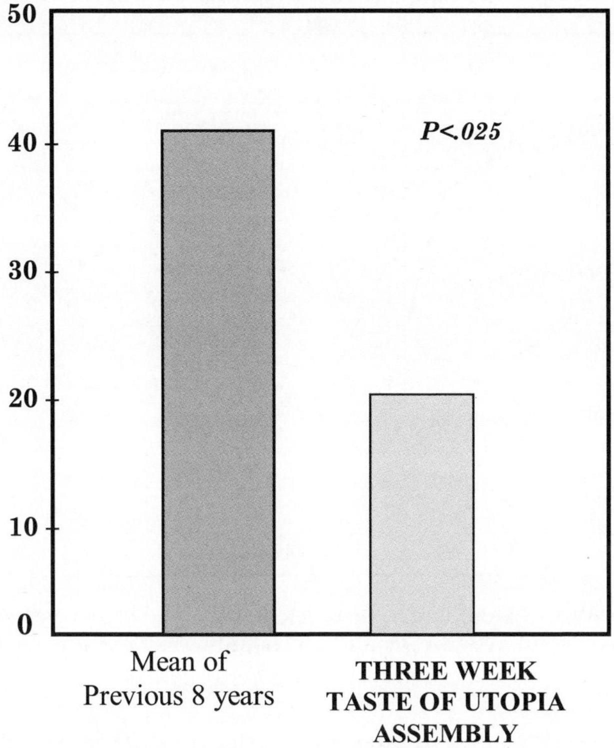

Chart 8 – *Decreased Traffic Fatalities per Miles Driven-USA*

Traffic Fatalities per Billion Miles Driven per Day over Christmas and New Year's Weekends

Traffic fatalities per day were at an all-time low over the Christmas and New Year's weekends during the Taste of the Utopia Assembly, even though miles driven per day were at an all-time high, eliminating the possibility that the decrease was a consequence of fewer miles driven due to cold weather.

Source: National Council, USA

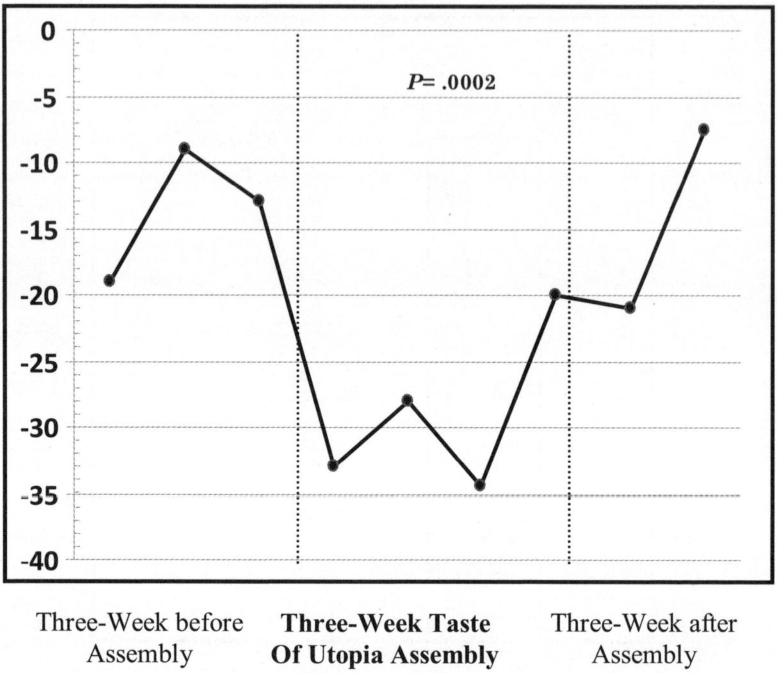

Chart 9 – *Decreased Incidence of Infectious Diseases– USA*

Percent Decrease in Total Incidence of Notifiable Infectious Diseases per Week Compared to Median of Same Week for Previous Five Years

The incidence of all categories of notifiable infectious diseases per week decreased during the Tastes of Utopia Assembly compared with the median of the previous five years for the same week of the year. After the Assembly the incidence of infectious diseases began to rise again towards levels comparable to previous years.

Source: Center for Disease Control, USA

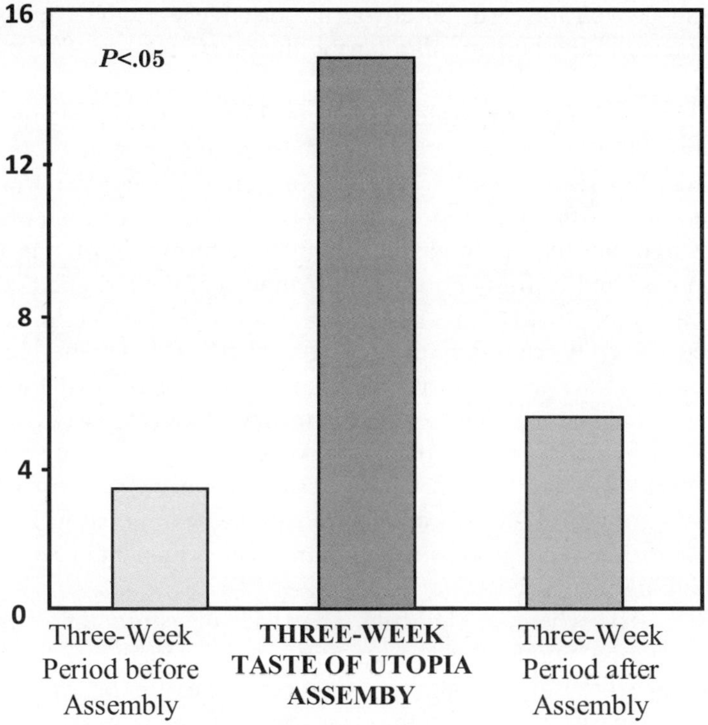

Chart 10 – *Increased Patent Applications in the USA*

Percent Increase in the actual Number of Patent Applications Compared to Expected Number

During the Taste of Utopia Assembly the number of patents increased 15% over the number of applications expected by the patents office for that time of the year. *After, the number of applications decreased to approximately the expected number.*

The results of the Global Taste of Utopia inspired Maharishi to establish heaven on earth based on the discovery of the Unified Field enlivened in the human mind during the practice of the TM and TM-Sidhi program in groups together.

Maharishi further repeated the event fifty times over and over documenting the events together with prominent scientists that established Maharishi's resolve that heaven on earth can be truly established on earth. The repeated events that illustrated similar results assured him that it is indisputable that the birthright of man is truly happiness and fulfillment bringing man to that highest status of life of heaven on earth. No one in the history of the world has proclaimed heaven can be created on earth. Maharishi was determined to create it based on the discovery of the Unified Field already validated and existing to hold the formula to create heaven on earth. Maharishi made a master plan to make it a reality.

Maharishi's description of the behavior and property of the elementary particles of nature can truly create heaven on earth. Like when the electrons and protons are contained in a light bulb and is switched on, light is produced.

Similarly, when the elementary particles of nature, the finest particles lingering within the human mind is contained in a large groups of nine thousand together in one place, heaven on earth is produced. Maharishi said that the Maharishi Effect is a scientific event. The Maharishi Effect is a discovery of science that needs to be considered by all leaders of the world. It is a strategy for global transformation much needed in our turbulent times.

7

SCIENCE BEHIND CLIMATE CHANGE

I first heard of the "sap" of the universe from Maharishi. He said, "It is like a tree, the "sap" permeates the root and nourishes every aspect of the tree." The "sap" of the universe permeates all creation in the universe including man. It is the unseen, unmanifest non-material root of all life. It is the Unified Field permeating the whole of life in the universe.

I correlated Maharishi's "sap" to the elementary particles of nature that permeates every living creation in the universe. Maharishi was a physicist and related the energy experienced in the practice of Transcendental Meditation to the Unified Field, home of the finest particles in nature.

If the particles at the Unified Field are the finest, finer than the proton, neutron and beyond subatomic particles, it must be the "sap" Maharishi referred to which creates the finest of nature's energy at the level of the Unified Field, the seed form of all energies in nature. Quantum physics describes energies in nature which contain the elementary particles like the photon particles in laser energy, the electron in the electric field, the magneto particles in magnetic energy or the graviton particles in gravity.

Quantum physics was not as well known more than fifty years ago, as it is known today. In college, in the early 1960's, classical physics was part of the engineering course. Thermodynamics is a field of physics which contained the background of particle science as it describes the transformation of matter with the application of variations in temperature as particles turns to liquid form, then to gas and converted to energy. Air, when warmed evaporates, when cooled turns to water, further cooled turns to solid ice.

Quantum physics refers to the product of the breaking of gas particles to finer forms and to the energy field they create. The finest particles referred to by Maharishi as the "sap" is described as the finest elementary particles produced and enlivened at the level of the Unified Field that Maharishi found to occur also at the most silent level of the human mind.

There are a handful of available books and websites about particles and their origins describing the elementary particles of nature and their corresponding character. There are several topics describing quantum physics,[17] some say fine particles have uncertain characters, that is why scientists formulated the uncertainty principle of particles. Particle science is most interesting as it is very strange. Since it is a new science, many physicists have many uncertain theories. Science has not yet found a name for the finest particles of nature or has it actually found its form. Therefore, for my reference, I called it as the AS_U particle, the anima solo particle that nourishes and animates the whole universe. It is the "sap" that Maharishi referred to that permeates all creation, it nourishes all life, it has its basis in the Unified Field. Maharishi said, "It creates by itself, from itself, within itself." That is why I referred to it as a solo particle.

Einstein sought for and never found the Unified Field, maybe because you have to do it, to find it. Maharishi said, "The practice of Transcendental Meditation is the "do it" procedure and allows the conscious mind to identify itself with the Unified Field in pure consciousness." Pure consciousness is found at the most quiet, deepest level of silence within the human mind. Finding what brings the good things that come from the deepest level of the mind is apparently as mind-boggling as searching and locating the Unified Field of all the laws of nature, like Albert Einstein persistently pursued throughout his life but has missed "to do it" and locate the Unified Field until the very end of his life.

Maharishi referred to the Unified Field as the "sap," and finding what the "sap" is made off, starts at the level of particle science, described in quantum physics. Just like the magneto particles create the magnetic field, the graviton particles create the gravitational field or the electron particle forms the electrical field in and around the earth, therefore, the ASu particle creates the Unified Field. So, the start of the quest for the answer to what consists Maharishi's "sap"' must start with the quest for the finest elementary particle of nature.

Maharishi said practicing the Transcendental Meditation is the easy thing and it is effortless, even a child of six years old can practice it and that is how you locate the Unified Field. Albert Einstein came from the West and had no knowledge of meditation, unlike Maharishi who learned Transcendental Meditation from his master, Guru Dev from the Himalayas where the traditions of the ancient Vedic civilization was passed on generation to generation after the ancient Vedic civilization of India faded away. The ancient

---

[17] *www.particleadventure.com*

Vedic literatures highly extolled the secrets of meditation and referred to it as "Brahm," in modern science it is called the Unified Field.

Scientists until now continually pursue the quest for particles and the most powerful physics machine ever built on earth, at the CERN in Switzerland. The Large Hadron Collider, located in Switzerland was built with the purpose of unraveling many unknown about the beginnings and functioning of the universe. In the most recent experiment conducted in July 2012, at the CERN, in Switzerland, the long predicted Higgs particle, a new particle that create mass to atoms was found to exist. At high speeds, atoms are made to collide and break down to finer particles. That way the most sought after Higgs particle, named after scientist Peter Higgs was found. Some call it the God particle. The discovery did not state that the Higgs particle was a solo or indivisible particle. The penta quark which intertwines with five quarks was later found after the discovery of the Higgs particle. The particle accelerator at the CERN in Switzerland have come up with hundreds of particles by colliding atoms at high speeds nearing the speed of light and break them down to finer particles.

Maharishi said that the Unified Field can be enlivened within the silent level of the human mind, it is the laboratory for validating its effects individually and in large groups together. The individual effect exponentially increase with groups together, like joining another magnet with groups of magnets, the magnetic field exponentially increases and becomes invincible from outside impurities. The Unified Field creates an invincible attribute which is exponentially more powerful in groups together that it cannot be penetrated by any negative trends, only positive trends occur.

I went thru many long nights reading and searching about the elementary particles of nature, the exciting adventure referring to the elementary particles of nature which were illustrated in interesting formats. It was a quest so interesting and exciting because particle science is a recent find among scientists.

Quantum physics barely started from the time of Albert Einstein in the early 1930's to recent discoveries to date. Einstein elaborated the nuclei of the atom. He identified the hydrogen particle and predicted the development of the hydrogen bomb. A concentration of the hydrogen particles when placed together in a closed container, in groups together created a sensational, powerful and destructive blast that wiped out a city in Japan during World War 11 in 1944. Maharishi said that the "sap," the Unified Field is the finest, purest particle in nature, it posses the nourishing character and does not contain any destructive property. Existing evidences has been made that displayed the nourishing character of the finest particle that creates the Unified Field..

Deeply curious about the spectacular effect of the group practice of TM and Yogic flying, I was driven to pursue the scientific basis relating to Transcendental Meditation which Maharishi repeatedly stated that TM is a science, a modern science that originated from the ancient Vedic traditions held by the ancient Vedic civilization of India. From the study of the basic particles in physics, to their complex finest form, it eventually made me feel secure about the ASu particle, the anima solo particle I referred to referring to Maharishi's "sap." Quantum mechanics illustrates the properties of atoms as they break down to smaller and finer particles until they split down to the finest of all particles, like breaking down neutron to neutrinos, boson particles to gravitons and so on until it is split down to finer quarks or other finer particles like the mesons.

Microscopic particles have strange behaviors. They are a particle as they are a wave and a string at the same time, a triad attribute. Beyond the atom, physicists refer to the Standard Model to describe the most recent smallest experimentally observed particles of matter.

Quantum theory refer to fields of matter, like particle fields of electricity or heat particles like photon fields or the magneto particles in the magnetic field. Particles of nature create and exist in a force field. The electric field, magnetic field, gravitational field and now Maharishi said the Unified Field is created by the "sap" formed from the finest particles of nature also lingering within the human mind at the level of silence.

Before 1984, quantum theory and Einstein's theory of relativity were entirely two different fields, until the string theory came about. From deeper underlying probed, the string theory found that both matter and energy belongs to one and the same field. Force fields and matter fields spring from the same source, the Unified Field. Particles brake down to finer and finer and become untraceable and unseen, like they are in many places in a split of a second. Fine particles join and fluctuate as pieces of strings instead! This is how the string theory came about.

Einstein's theory of relativity described the force of gravity and the relationship to the energy field produced on each particle or on each planet. It is like the earth's hovering magnetic field or the gravitational field that keeps all the planets in place without falling out of their orbit, like the earth moving perfectly around the sun along its orbit.

There are categories of particles that form the Standard Model. Scientists sometimes call it the M-Theory - master, maker or miracle particle theory. This "out-of-nothing to something" theory is like Maharishi's "by itself, within itself and from itself" or "sap" theory from "out of nothing to something."

Similarly, the "out-of-something to nothing," is a theory likened to the process of introducing heat to solid matter or solid ice-something, in matter form and when solid ice is warmed it turns to liquid water. Heat it further to boil and the liquid water turn to unseen air particles appearing to be nothing. Or vice-versa, "from nothing-to-something," like air particles appearing "nothing," then turning to liquid water when cooled at lower temperature, then further cooled to "something"- solid ice, matter. The three states of water contain the same molecular content, in solid ice, liquid water and unseen air. They contain the same hydrogen and oxygen particles.

I imagined Maharishi saying, "From zero to diversity in nature. He would repeat "Zero.., Zero...," curling his lips to form like a big zero. He played with the word zero. It just kept me in total wonder for a long, long time, wondering what he was referring to. Ice becomes matter at below zero degrees centigrade. This "something" state maintains the same particle or molecular content as water turns to air. Further, these hydrogen and oxygen particles break down to finer unseen elementary particles, to the "nothing" state, like unseen air particles. They become the state of unseen matter, the "nothing" state. They appear to be nothing because you cannot see these tiny particles. We breathe the unseen air particles or we die without them. They exist, these fine unseen particles!

Matter particles are composed of fermions made of quarks and leptons; energy field particles consist of bosons carrying the three forces of nature, electromagnetism, the strong nuclear force and the weak nuclear force. Energy is produced by the breakdown of unseen matter, when they break down to finer then to the finest particles. Like in the physics law of thermodynamics describes the pressure or energy of gas particles contained in form. The smaller and finer particles become, the greater the group energy they produce. Power of the energy fields are described in the Planck scale in physics illustrated in the Maharishi Physics Chart on next Chapter 8. Finer to finest form, they bring about exponentially more powerful energy field. Particles produce energy force fields like electrical energy with electron and proton particles, magnetic energy with magneto particles and gravitational energy fields with graviton particles. They act as minute unseen strings vibrating in space. The string theory is a candidate for an effective and complete theory of the beginning of the universe that Einstein was searching for and never found.

Scientists now call the string theory, the theory of everything where nothing apparent and visible seems to exist! Maharishi's "sap" from "itself, by itself, within itself" created something out of nothing appears to be the theory of the Creator! Like the ASu particle is the particle of creation. This string theory make heads spin and probably will not at this time understand their indescribable, uncertain and strange particle behaviors. It has overwhelmed many scientists and referred to it as the uncertainty principle.

How could these uncertain particle behaviors be related to Maharishi "sap" or to his Transcendental Meditation program and why the benefits are so profound? Where do all the positive effects of TM come from? How does the Maharishi Effect be so powerful as to eliminate negativities on earth? How could the Global Taste of Utopia display such spectacular positivity to have affected the whole world?

Many scientific researches have been done and said that the Maharishi Effect has a scientific basis, but I was curious for further data to find out where the scientific basis of TM and the Maharishi Effect is made of. I know the Unified Field creates the Maharishi Effect, but what is it that forms the character of the particles that creates it, like the character of invincibility or the cooling and nourishing character of the finest particle at the level of the Unified Field. There are characters of natural law explicitly observed, like a balloon rises when filled with air or snow melts with heat. The human body has natural characters designed by nature, like as soon as a cut bleeds, blood clots, or when warmed, the human body activates its sweat glands furthering the behavior and character of water, when cooled, air turns liquid. When water is heated to 100 degrees centigrade turns gas, further cooled at zero degrees centigrade turns solid ice, these are the three states of water, liquid water, ice as solid matter and gas as air, yet maintains its constitution, H_2O. The observed water droplets from the skin when warmed illustrate the natural cooling mechanism of the human body. Are observations enough answer to the questions why they do what they do? When gathered in large groups together it can cool the atmosphere in an instant and solve global warming! I was not alone in the observations. In Mozambique in 1994, as soon as large groups together in the military of Mozambique practiced TM and Yogic flying together, rain fell in an instant in areas where drought prevailed for more than twenty years. From December 19, 1983 to January 6, 1984 in the Global Taste of Utopia, rain fell in an instant in drought stricken Brazil, when the threshold of 7,000 was reached, it was the square root of one percent of the population of the world then in 1984. Maharishi promised the people of Brazil that rain would come and it did. How does the Unified Field melt down all the negativities in the environment? In classical physics, the electric field burns any impurities that come into the electric field. Is the character displayed by natural law enough answer? I wanted to know why and what causes the Unified Field to move negativities out of human life and what causes heaven to come down to earth. Maharishi said that the Unified Field creates heaven on earth. I wanted to know what Maharishi's master plan to create heaven on earth is made of. If human suffering can be made to disappear, truly it is a great endeavor to search for. The idea of heaven on earth is just as exciting as removing suffering and problems as Maharishi has foretold. Religions preached heaven after death. Maharishi laughs and says, "Now, not after, heaven can be created by the Unified Field on earth for man to enjoy now, not after death!"

My curiosity was intent, about heaven being created on earth. I diligently practice TM without missing and learned Yogic flying. Maharishi said that these practice show where the Unified Field is found. TM is the "do it" procedure. I continued reading and researching on particle science to find out the reason for all the good things happening with it, from the practice of TM and Yogic flying! Reading and researching was not enough. Maharishi's instruction was, "Do it. Just do it," he said. Maharishi said, you have to do it, to find it! That was exactly what I did. I formed the group and observed the same results Maharishi has foretold and I found the answer!

Maharishi stipulated that uniting gravity with the three known forces, electro-magnetism, weak force and strong force do exist and their point of unification ultimately exist in pure consciousness. He said that the junction point of unity occurs within the human mind. The Maharishi Effect is produced when large groups are gathered in one place together and practice TM and Yogic flying, it creates Unified Field at the level of the quiet mind. Why it has to be quiet? Why it has to be in groups?

Fine particles of nature behave exponentially greater when concentrated in form, like one photon particle produce less heat than when in groups contained together creates the powerful laser. The finer particles become, the field they create becomes exponentially greater. It is their natural character. To prevent entropy and create maximum effect, they have to move in a quiet, undisturbed and coherent environment. The human brain attains maximum coherence during Yogic flying, and at the level of silence, the body lifts up defying gravity.

The Standard Model refers to atoms. It is incomplete and does not include the fourth force, gravity. Physicists find gravity illusive. Even Einstein could not confirm the unity of gravity with the other three forces, electromagnetism, weak force and strong force. These forces are results of specific and defined force fields made by recent discoveries of science produced by particles of nature as they transform to finer forms. Temperature change is normally the catalyst in the transformation.

Beyond the Middle Age, philosophers found that there were only three basic particles and categorized atomic particles from electron, neutron and the proton. Then early in the 1930's, neutrons were further found to split finer to neutrinos, fermions and quarks. Bosons break down to photons, other particles to gravitons, W or Z, gluon or muon particles and pi mesons.

Michio Kaku[18], a contemporary physicist stated, "As an example, look at your body. The electrons and protons that make up your body are all fermions. Everything you see around you, including the walls and the sky is made of fermions..."

Then I began to understand Maharishi's "sap." These particles also exist in our bodies! Our body and mind consists too of these tiny particles! No wonder, at the silent level of the mind, man produce the predicted coherent behavior of the finest "sap" or the finest elementary particles of nature at the level of silence during the practice of Maharishi's Transcendental Meditation program. It is physics discovered in the human brain! It is science of particles found within the human brain! They create the Unified Field!

Physicist Stephen Hawking[19] another contemporary scientist concluded that "....we human beings, who are ourselves a collection of the fundamental particles of nature - have been able to come up close to understanding the laws governing us and our universe is a great triumph."

Maharishi had described the finest particle of nature in a more simple, layman's terminology as the "sap" of the universe. Our body that includes our mind is the container of the finest particles of

---

[18] *Beyond Einstein, The Cosmic Quest for the Theory of the Universe, 1995, Michio Kaku and Jennifer Thompson*

[19] *The Grand Design, 2010 Stephen Hawking and Leonard Mlodlinow*

nature! These particles linger within our brain! The finest unseen particles break down to smaller particles. Finally it becomes the indivisible root particle of all the energy particles in nature. I called it the lone particle, the soul or solo particle or better still, the anima solo particle, the AS_U , the solo particle that nourishes and animates the whole of the universe.

The solo particle appears to be indivisible and does not need any other particle to create its own energy as Maharishi said "from itself, by itself, within itself." Not like in the electric particle, the electron, it needs a proton and two quark particles to create the neutron particle. The water molecule needs two parts hydrogen particle and one part oxygen particle to create the water particle. Some physicists say particles come in pairs. They come with an anti-particle.

In 1937, an Italian physicist, Ettore Majorana predicted that there is a single fundamental particle, and it is at the same time its own anti-particle. Majorana disappeared during a trip to Palermo, Italy. He never came back home and was not able to prove his prediction.

Another physicist, Vincent Mourikand Leo P. Kouwenhonen states that this singular particle that Majorana referred to, can exist by exposing a small circuit to a magnetic field. Kouwenhonen still needs to confirm his findings.

Physicist, Julian Schwinger once said, "There are a heck of a lot of energies we haven't gotten to yet." There have been more than two hundred particles experimentally discovered to date that are unseen by the naked eye. Maharishi referred to the energy field created by the finest, indivisible particle as the Maharishi Technology of the Unified Field. It is a field where all energies of nature unite as one seed form. See unification of all the forces of nature in the following Chapter 8 in The Discovery of the Unified Field.

Some call this finest particle, the deity, the God particle or the particle of everything. Maharishi called it the "sap." It is the "sap" particle! Like a sap inside a tree running from its roots to it branches and nourishes every part of the tree, the leaves turn green, the flowers blossom and it bears fruits! It is inside all of creation, inside our body and mind, inside the walls or the sky above.

The "sap" that Maharishi referred to appears to be the solo particle. Or it is the particle of everything that is by itself the finest particle that creates from itself all the other energy particles currently known. It could have been the Big Bang particle that scientist referred to that blasted off the origin of the universe.

Maharishi called the field where all the tiniest, indivisible solo particles exist, the Unified Field. He called it the "sap" that exists and creates "from itself, by itself, within itself." He said, "the knower, the process of knowing and the known" value of the universe (whatever that means). Maharishi stipulated that it is enlivened and can be drawn at the silent level of the human mind. It becomes alive during TM! The finest particle of nature exist in our brain! Atoms and particles become unstable even with the slightest disturbance. That is why Maharishi said, "The mind has to be at a state of silence."

Maharishi's Transcendental Meditation quiets the mind. To be in-phase and coherent, fine particles need to move in a quiet undisturbed brain environment. TM quiets the mind where the solo particle lingers. The elementary particles of nature create a field, like the magnetic field, electric field and gravitational field. The anima solo particle, the AS_U particle I referred to creates the Unified Field! It is the finest particle that creates "from itself, by itself, within itself!" If it is the finest, it must create the most powerful field because scientists say that the finer particles become, the more powerful field they create. So it becomes invincible from outside interference, like the electric field, it burns all other strange inserts in its field. The magnetic field prevents any outer influence to penetrate its field. Like the magnetic field of the earth, it shields the earth from solar flares! Excessive sun's rays cannot penetrate the magnetic field of the earth! It is a natural character of the magneto particles.

The Unified Field is the purest field in nature, so no negative trends can penetrate its field, that is why the Maharishi Effect is so powerful that only positivity exists and no negative trends can insert in its field - in the invincible Unified Field! This describes natural law principle of invincibility. The field created is invincible from negative trends. It is the purest field in nature, thus only positive trends can exist. Positivity is its natural character. Invincibility means inability to penetrate its field, or break the shield it has created.

Like the CO_2 gas cannot penetrate the earth's magnetic field or the gravitational field. That is why the CO_2 is trapped above the earth's atmosphere below the magnetic field, it cannot penetrate the magnetic field or the gravitational field of the earth and disperse into space, that is why it stays above the earth's atmosphere and absorbs heat reflected from the earth and warms the earth's atmosphere, it evaporates the air at the ozone layer that protects the earth and cause droughts and global warming. In the cold season, it absorbs the warm air from the ocean, cools the air, producing massive rain and brings floods and rapid wind turbulence on earth.

The graviton and magneto particles are much finer than the CO_2 particles, the finer particles become, they possess the invincibility character of natural law. The finest particle of nature, the AS_U particle, the primal particle of nature contains the finest and purest particle and it becomes the most powerful of all particles that creates "from itself, by itself, within itself" and creates the invincible Unified Field.

Hydrogen particles cannot penetrate the earth's magnetic field because the magneto particle is finer than the hydrogen article, like the hydrogen bomb that was dropped in Hiroshima, in Japan in 1944, the hydrogen smoke rose from the earth to the atmosphere above, but as soon as the magnetic field was reached, the hydrogen particles curled down forming a mushroom cloud above the earth's atmosphere. The hydrogen particles did not shoot up and disperse into space, it cannot penetrate the earth's magnetic field.

The same happened when the Pinatubo volcano erupted in the Philippines in June 15, 1991. It was the second largest volcano eruption in the twentieth century. It spewed columns of ash miles up into the earth's atmosphere. It shoot up but did not disperse into space as the volcanic gas is made of CO_2, it cannot penetrate the earth's magnetic field, instead, it formed a cloud of gas and spread like an umbrella hovering above the earth. CO_2 is soluble in water. Rain fell as soon as the air in the atmosphere cooled off, bringing the volcanic ashes down to earth. If the rain did not come, the CO_2 layer formed would have stayed longer above the earth, it would absorb more heat from the earth and cause more global warming or when cooled it will cool the warm air from the seas causing floods on earth.

The ASu particles in the Unified Field is the finest, finer than the magneto or graviton particles, so it can penetrate the magnetic and gravitational field and disperse the CO_2 gas into space and thus eliminates global warming and drastic climate changes. The Unified Field is the most invincible shield that can be formed on earth just by establishing TM and Yogic flying participants in large groups together in any one place on earth.

The process of creating the invincible Unified Field is simple enough for any government or groups of wealthy individuals to implement, just by gathering the square root of one percent of the country's population. For the world, 9,000 is the number needed at this time in groups together to quiet their minds in one place. All governments have military personnel already in place that can do it and find the solution to climate change and solve any negativity prevailing on earth. The cost is less, much less than the $10 Billion dollars budget proposed by the international groups monitoring climate change or the $100 Billion dollars they proposed for the year 2030. No negativity can exist within the Unified Field. That is why Maharishi said, "It is now possible to create heaven on earth."

There are groups in Fairfield, Iowa, USA at the Maharishi University of Management, where students and Fairfield residents join together at the Golden Domes built at the university campus that has been going on for more than forty years. It had more than 1,000 participants when the Vedic pundits from India were in Fairfield, but now many pundits have gone back to India.

Maharishi has established the Maharishi University of Management in 1974, it was then the Maharishi International University. Maharishi's intent was to grow the roots of TM in the USA in order to support the USA in its role of fostering peace and expanding economic prosperity in the world. At the same time the Maharishi Vedic City was established, also in Fairfield, Iowa and brought Vedic pundits from India with Maharishi's intent of creating large groups in the USA, about 2,500 participants are needed to make the USA truly an invincible country.

Maharishi founded the Natural Law Party in 1992 and made Dr. John Hagelin run for the President of the United States but did not garner sufficient enough popular response from the population. In 1993, Maharishi again organized a two week assembly of a group of 4,000 TM and Yogic flying participants to gather in Washington D.C and once more to convince the USA government that Maharishi's Unified Field principle can indeed provide unprecedented and great benefits for the USA. Washington D.C was then the crime capital of the country with about five killed per day occurring at that time in 1994. The assembly reduced crime rates down to 23% in just two weeks and the stock market rose as Maharishi has predicted. Finally, Maharishi left the United States and disbanded the Purusha and Mother Divine group, about 1,000 of them, all practitioners of TM and Yogic flying group who mostly worked and followed Maharishi.

This time he proposed the establishment of the group in India. Maharishi took the responsibility to establish the 9,000 group for the world, currently, there are nearly 1,000 Vedic pundits at the central part of India, called the Vedic pundits at the Brahmasthan of India (See vedicpundits.org). The 9,000 has not yet been attained to date.

Meantime, ferocious and massive wild fires raze in California, Nevada, Washington State and other states destroying infrastructures and thousands of homes, multitudes of individuals are displaced, along the massive lines of fires, many are hurt and some tragically killed including firemen, at the same time massive storms have flooded Louisiana, New Jersey, Texas, Florida, Colorado, Utah and other states. There are billions of dollars in properties and infrastructures that fires have burned or that were washed away by many flash floods.

Trillions of dollars are being spent by the US government in countering terror and conflicts around the world, meantime, also providing huge amount of dollars maintaining personnel and military bases all over whole world. Air strikes in Syria cost millions of dollars per day, degrading ISIS has been described by the West as a long process, and may carry a long string of costs, meanwhile the Syrian people are suffering for who knows for how long, already millions have fled war torn Syria.

In Mozambique, Africa, as soon as the Unified Field principle was applied, the rebels laid down their arms. The same occurred in my fourth TM school in a remote poor town in the Philippines, the rebels vanished as soon as the five hundred students started to practice of Transcendental Meditation in groups together. The same results occur each time the same Unified Field principle is repeated!

Maharishi has proposed a budget of just one billion dollars to feed and maintain a group of 9,000 and it is just a pittance compared to the large amount being spent by the US government including providing aid to many countries or much less cost building a four billion dollar bomber plane. It would even cost nothing using a group of military who are already stationed in several places, at home and overseas, all draw salaries, just 9,000 for the world participating in Maharishi's TM programs twice a day, morning and afternoon, before work and after, it is sufficient to prevent problems arising. It would be less in number if formed based on the population of any country beset with problems.

Maharishi's invincible principle of natural law can solve global warming surprisingly in an instant, counters conflicts and fosters prosperity for the world. All negativities in the world can be prevented by just gathering 9,000 in groups together at the level of the Unified Field. It is like the "sap" inside the tree, just water the root, the whole tree gets nourished. The Unified Field waters the root that nourishes the whole planet.

Maharishi made the Unified Field effect experiment in the documented Maharishi Effect done in 1983 described in the previous Chapter 6, in the Global Taste of Utopia which illustrated the positivity that was created around the world with the square root of one percent of the world's population, just 7,000 then participating in

TM and Yogic flying in large groups together. Afterwards, more than fifty events followed that further validated his discovery. I made experiments with the Maharishi Effect in a small scale and it similarly worked as I described the results of the events. The events were formed on a temporary basis.

Unseen particles act as waves, like light waves and sound waves. Light particles and sound particles move constantly in space. They are particles that possess the property of a wave at the same time. Like a school of fishes moving as a group in the ocean, one particle fish is as visible as the school of fishes moving together. Atoms and particles in nature produce a similar wave effect. One atom alone is insignificant compared to the effect of a group of particles contained and moving together.

One fish moving alone produce an insignificant wave than the group effect of many large groups of fishes moving together. Group minds silent together in one place produce the spectacular Maharishi Effect! Movement of the fine particles in nature are unseen in space. Scientists refer to moving wave particles as the string theory. Fine particles act as waves and string at the same time.

I began to understand the basis of my anima solo particle and its behavior producing positivity in the Maharishi Effect theory. I became truly excited to discuss the energy field effects of this finest particle of nature occurring within the human mind to a close friend one time. He never encountered the term anima solo particle in his life time, I believe no one has, it is the terminology I used to define this particle, naturally, he seemed uninterested.

"I never heard of this anima solo particle you are talking about. No one knows about it, I am sure," he said. "You can talk to me about it forever, but I am too old to understand and benefit from it," he continued with sarcasm.

"You can read the spectacular effects of the finest particle within the human mind. I discussed them at length," I insisted with excitement. "You might miss the greatest discovery of modern physics to date. It can extend longevity and reverse the effects of the aging process. It can even completely eliminate disease and human suffering. It can create heaven on earth!" My friend stared at me wide-eyed perplexed about my "heaven on earth" statement.

I continued, "The significance of understanding the finest particle in nature is that the human mind and body becomes the conductor of the M-particle theory referred to by scientists," I concluded. "Or the God particle many scientists refer to."

One lady acquaintance read my book. She said, "I could not stop when I started reading it. It was very interesting." She told me. She was a seeker of knowledge, that's why.

"I have been going from one church to another church. I did not find any substance to make one happy from any group," she said. I said, "TM is not a religion. It is a procedure that keeps the mind stay quiet. The regular practice of TM makes you happy afterwards. You tap that bliss particle found within the human mind."

The M-particle theory maybe interpreted as the "master, miracle or maker" particle theory. I call it the "anima solo" particle, the seed form of nature referred to as the "sap" by Maharishi in his technology of the Unified Field.

Metals are the conductor heat, like the photon heat particles. They are heat particles in lasers. The copper wire is the conductor of the electron particles producing electricity. The magneto particles in magnets are the conductor of the magnetic energy.

Man's mind at its most silent state produce the Unified Field that scientists are seeking for! The human nervous system is the conductor of the Unified Field! What an exciting and significant discovery eliminating all negativities in the life of man on earth!

I consider it the highest privilege for man to be in that status of being the conductor of the total potential of natural law at that M-particle theory category. For man to be the channel of the singular, finest and primal energy of the universe, above all of the particle energy carriers of nature, the "master, miracle or maker" category, whatever name it may be called, it is indeed the most supreme of all knowledge as Maharishi described it to be!

It is the greatest triumph in understanding the laws of nature governing man and the universe. It could be the "out-of-nothing to something" particle theory of the universe. At the silent level of the human mind, gained thru the practice of Transcendental Meditation in large groups together, man becomes master of natural law. It is truly the greatest triumph for man to finally discover his birth right, to gain the right to live in fullness beyond suffering. This knowledge is of great importance to everyone. One unseen, unknown particle becoming the master mind particle of the universe!

I think that man was created for this most supreme and ultimate purpose. Now I understand, it all boils down to science. TM and the Maharishi Effect truly has a scientific basis. Science describes the Maharishi Effect occurring when big groups practice TM, quieting their mind together in one place producing an enormously awesome and spectacular effect that uplifts the well being of humanity to the highest status participating with the functioning of natural law.

Observation is the basis of science. If the observed result is the same in every event or experiment, it becomes a fact. Maharishi called for an experiment, observed and documented. It was written in the previous chapter 6, The Global Taste of Utopia. It was one of the more than fifty events that repeatedly produced the same observed results. It is similar to the effect of a bigger copper wire conducting greater amount of electricity. Bigger or more concentrated photon particles together produce greater heat described in laser technology.

Similarly, the Miessner effect in magnets illustrates a phenomenon of science, which describes the effect of bigger magnets joined together creates a larger magnetic field. The greater the field that is produced, the greater it eliminates impurities within its field.

Award winning physicist, John Hagelin, PhD, confirmed Maharishi's theory of the Unified Field uniting all the four fundamental forces of nature occurring in pure consciousness. Dr. Hagelin was awarded the prestigious Kilby award placing him at a similar high level of status among physicists like Bohr and Eddington. Dr. Hagelin worked extensively with Maharishi during his lifetime. Both physicist, Hagelin and Maharishi came up with the Physics Chart (shown in Chapter 8) formulating the relationship between matter and energy fields.

The illustration of the unification of the four fundamental forces of nature made by Dr. John Hagelin under Maharishi's guidance is described in the following Chapter 8. Maharishi made Dr. John Hagelin, the National Director of the TM movement for the whole of the United States which he still holds today.

Maharishi passed away quietly on February 5, 2008. He was 91 years old when he died. I remember when my father died in 1967, a plate from a nearby table fell and broke into pieces the very time and day my father passed away. The same strange thing happened exactly the same time and date Maharishi passed away. He was my teacher. I hesitated to mention the incident to a friend.

I was surprised he agreed with me and said, "You know I stayed in the mountains sometime in my life and there were elders in the mountains who said the same thing happens when someone close to you dies. Their spirit particle passes through some unseen cosmic field and sends some message like this at the time and moment of their death."

I said, "It might have been just coincidental, but this happened twice, my father and Maharishi, very strange." I was very close to my father. Everyone in my family knew how close I was to him. Because of my TM school projects, I became close to Maharishi like he was my father, so few new about.

All particles of nature are interconnected, like a message from the International Space Station flowing at lightning speed through space and received on earth. Or the modern email received worldwide in seconds of time. They operate on the principle of interconnectivity of particles of nature, electric particles, magnetic or the same particle field in wave form used in communications.

After I learned TM, I wanted my children to learn it too. My children were too important to me to make them practice some kind of an untested science. I wanted to validate Maharishi's theory first because I wanted to be sure TM is a true science. I found TM not only is a true science. As Maharishi said, it is the most supreme of all sciences being the seed form of natural law.

There are many books written about Maharishi. The story here is my personal understanding and interpretation of the incredible findings about the Maharishi Effect. I had the opportunity to experience and validate myself.

I was in India at one time to see Maharishi. I asked him, "I should write a book about you, Maharishi, so people would know more about your awesome teachings." I sat on my knees on the white marble floor looking up at him sitting on the couch. "Maybe, I should have a picture with you, so people will know, that I wrote the book with your permission." I faintly suggested to him.

Our picture is in the following page. "The floor is cold there. Sit here," Maharishi said, pointing to a chair nearby. He instructed the lady who came in with me in the room. "Take a camera." After the lady left the room, he instructed me. "Gather all the data you need and write." Then, Maharishi started to draw some sketches on the yellow pad paper on the small table in front of him. I bent my head curiously towards the paper he was writing on, looking at it as he started to draw.

"See, the mind is like an ocean," he said drawing the waves on the yellow paper. "At the surface level, there are many waves. As you go down to the bottom, it becomes more quiet," he repeated, "more quiet until you reach the quietest bottom of the ocean where there are no more waves." My attention wandered for a while away from Maharishi. My mind was fleeting to the lady I came in with, who took our picture and left. I started to wonder if she was able to get the taxi we needed to get to the airport to go back home.

Maharishi continued drawing. "This is similar with the mind. When you close your eyes, and reach the quiet level, where there are no more thoughts, beyond thoughts," Maharishi continued. "Quieting the mind is very important. I consider my Transcendental Meditation a mental technique that is very effective in quieting the mind."

Maharishi looked very serious and solemn-faced. He appeared unmoving sitting in a lotus position on the couch. I focused my attention back to what he was going to say next. Maharishi was short and had a small frame. From the chair I sat down after rising from the marble floor, I towered over him. That was why I sat down on the floor when I came in.

"See the bubble," Maharishi continued. "From the quiet ocean floor to the surface, you become aware of the meaning of your thoughts on the surface level," pointing to the bottom of the ocean on the paper he drew on. Maharishi's Ocean diagram is attached next page. "This is what we need to teach people to do, quiet their minds. My meditation can quiet the mind easily without effort. "See how simple it is," pointing to the diagram he made on the paper. The simplicity of the diagram contrasts with the profound wisdom that is at the bottom quietest level of the mind and its relationship to the finest elementary particles of nature described in quantum physics. Maharishi wanted his message understood by everyone and illustrated his principle in the most simple and understandable diagram.

"If you are not meditating, you use only five to ten percent of your mind. You use the maximum mental potential, one hundred percent of your mind when you meditate," Maharishi continued. The diagram looks simple enough. Maharishi said that the profound knowledge beyond the deep level of the mind far deeper beyond the surface is most vital in the life of man.

I began to understand what he meant after the exiting particle adventure I had read. One physicist, Lacomte Du Nouy[20] said, "The splendor of the world was born from the impact between the universe and our consciousness." It is like what Maharishi said, "Consciousness at the level of silence of the mind is the spring board of creation. It is the seat of natural law." The meaning of what Maharishi was illustrating in his ocean diagram kept me wondering for some time. The simplicity of the diagram and the spectacular knowledge beyond the level of thought appeared very strange in the beginning. After a little while, Maharishi was quiet. I was silent too, eager to hear more from him. He remained silent. I broke up the

---

[20] *Human Destiny, 1962, Lacomte Du Nouy*

seemingly solemn silent moment. I stood up quietly and said, "Maharishi, my taxi is waiting and I have a scheduled flight back home," I said pointing to my watch. He did not say anything. He just looked up at me. He was very sad.

Then he said, "We should put up a school here in India. You should stay here in India and help with the school." I shook my head and answered him, "Maharishi, I have three small children. I cannot leave them alone without a mother. I will prepare the school curriculum for you, the same ones I use in my school," I promised.

It was impossible for me to leave my small children alone at that time and work with Maharishi in India. I would never have left my children alone for any reason. I stood up and walked slowly towards the door. I looked back at him before I left the room. He was still sitting there on the couch looking sadly back at me. I was in a hurry. I could not have left him alone in the room if I did not have a plane to catch. I left the door open, as I hesitantly moved away.

I was in the Netherlands too, one time, he said, "I was told that you have already left." I answered, "Maharishi, I am still here." I was always in a hurry back then, with so many concerns. The issues about my three small children, the school and other engineering business I was into then took so much of my time and attention. It was very difficult for me to sit down and listen to Maharishi for long hours. We were always listening to him for hours at a time. I prepared and gave him the school curriculum he was asking for.

Maharishi's teaching spanned over fifty years and captured nearly six million people worldwide who learned his procedure of Transcendental Meditation. The bubble diagram has become the basic illustration for understanding his system of meditation known worldwide as the Transcendental Meditation technique. It was the same message he repeated again and again in his travels all over the world with the same bubble diagram of the ocean. The same orderly government of nature operating in the whole universe is the same government functioning within the human mind and body. "As above, so below," many ancients say the same. The human mind is part and parcel of the constitution of the universe. Man can harness the most fundamental orderly functioning of nature within his mind thru Transcendental Meditation technique practiced twenty minutes twice a day morning and evening. At that quiet level of the mind, you are one with nature and the universe.

When the human mind attains that most quiet, silent state, man's nervous system habituates itself in that unique state, then the

Author Erlinda with Maharishi taken in 1988

mind begins to tap into the main switchboard of natural law, at the seed form of nature. The governing intelligence of nature takes over your life. You then begin to live life in accord with nature that is problem-free.That quiet government within man is mistake-free. It is the government of nature in the universe that is infinite and orderly All governments should utilize TM and govern in accord with nature. It makes running a government easy. Man and man-made governments accumulate problems for failing to tap into the Unified Field, this government of nature within man that is perfect. When man violates natural law man starts to have problems. See the planets in the universe, they move in perfect order. When we place our mind into that silent level, at the Unified Field, we tap that silent government of nature, man begins to live in accord with all the laws of nature within that is orderly and nourishing. It is in that quiet level of the mind that man gets in touch with the infinite functioning of nature in the whole universe.

Most of the times when we are awake, our mind is not quiet, always thinking this, always thinking that. Transcendental Meditation quiets your mind. It is when you quiet the mind that you tap into the quiet functioning and governing level of nature. Silence in the mind connects man to the silent state where nature operates at the level of the Unified Field. Silence in the mind is the most supreme state. The Unified Field is the main switchboard of natural law.

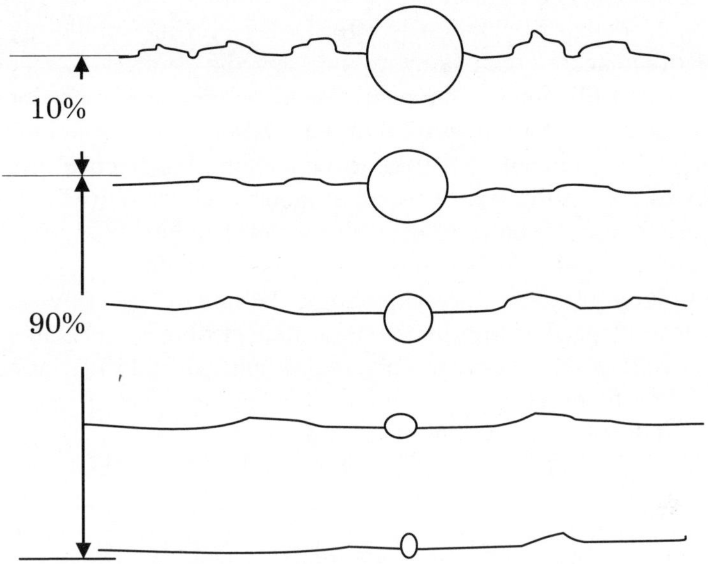

MAHARISHI'S OCEAN DIAGRAM

The silent administration of natural law is without a noise. Silence, zero activity is the basis of all diversity in nature. It is not a religion. It is a mental procedure. We do not pray. We just keep our mind to stay silent. When one prays, the mind is active. Silence, inactivity of the mind is the key. All types of government, either democratic, parliamentary, communism and whatever type any country has, there are always problems sprouting to the individual and to the collective interest of its people. Always, we see conflicts arising in government administration and in the surroundings, problems occur every day. With TM and Yogic flying in large groups together formed in one place, the surroundings become problem-free.

There are always some problems popping up somewhere in health, education, economy, crimes, wars, droughts, typhoons, people protests, earthquakes and name it, there is always a problem. To be problem-free is to have big groups that keep their minds silent together in one place. The government of nature is problem - free.

I remember Maharishi lecturing for hours and repeating the same concept of administration thru natural law. His concept of the constitution of the universe utilizes the laws of nature occurring within the human mind. The government of nature is a state of perfection. The administration of natural law is effortless. Transcendental Meditation is effortless. Everyone should take it and live life in accord with nature that is easy and effortless.

I truly found administration of natural law most interesting correlating it to administration of national government and to the well being of the citizens. I tried to understand the science of quantum physics that followed Einstein's theory of relativity and how it described the attributes of nature's particles and the energy fields they produce. I diligently explored the topic of quantum physics. It was truly interesting as it was strange. Eventually it provided me a glimpse of particles properties and their attributes. I related it with what Maharishi was saying about the Unified Field that administers the silent functioning of the universe including the earth and man thru the intelligence and design of natural law.

For more than fifty years, Maharishi made thousands of quotes and speeches that have been taped and archived for viewing. He expounded and repeated again and again, in his lifetime, the same benefits of Transcendental Meditation and its effects-improving health, expanding the mind, improving behavior, nourishing the whole society and balancing temperatures on planet earth. Maharishi described TM: "During Transcendental Meditation, as the mind proceeds to experience finer states of thought, the whole body becomes calm and quiet. This quieting of the body naturally allows an unusual degree of rest, which itself stores energy to a considerable degree. It is certain that the activity of the mind and the nervous system is least in this state, and, thus, the mind becomes quiet and calm. In this quiet state the mind and nervous system are alertly poised like the arrow drawn fully back upon the bow; they are without activity, but the entire system is alert in stillness. At the same time, all the mechanisms of the body are acutely balanced and steady. This restful alertness of the nervous system that is the most healthy state and is the basis of all energy and action."[21]

There are more than six hundred scientific researches documented by several prestigious medical journals worldwide. The prestigious National Institute of Health of the United States has provided some $24 million grants for some of its research. I have repeated the same notes that Maharishi had said too, remembering what a friend told me one time. "Keep repeating what you believe is

---

[21] *Transcendental Meditation: Science of Being and the Art of Living, Maharishi , 1963*

good for man. Many men never open their eyes to the truths of life, like a kitten it takes two weeks for its eyes to open after birth. A puppy, in one week, his eyes open up. Some men never open their eyes until they die. They are born, grow, mate and die. Repeat what you think can help others." Maharishi repeated the same for more than fifty years. He finally said, "We have planted the seed."

The Reversal of Aging Charts (See next page) has been printed again and again shown in scientific research found on tm.org. It shows how the reversal of the biological age happen resulting from the deep rest experienced during meditation. The body gains deep rest twice deeper than eight hours of sleep in just twenty minutes while the mind is alert. Growing old seems to have a reverse gear system towards growing younger. If you have meditated regularly and consistently for ten years your biological age shows ten years less. Mental fatigue is one cause of aging just as excessive physical activity. Blood pressure and insomnia goes up as you age, both go down with meditation. TM reverses the aging process. It has been documented as a major stress-buster that supports health and well-being. TM is vital in alternative healing and preventive medicine. There are testimonies from several medical doctors documented describing the health benefits of TM. The benefits of Maharishi's system of meditation on the individual level has been validated and documented in many medical journals (See tm.org).

The group effect described in the Maharishi Effect was experimented more than 50 times and currently there is one in India described s the Vedic pundits at the Brahmasthan of India. It is most overwhelming to note Maharishi's correlation of his meditation to the "sap" or to the finest particles of nature describing how the four fundamental forces of nature, electro-magnetism, weak force, strong force, and gravity unite as one unified energy that is enlivened within the human mind at the level of silence.

The unification of all these forces, described here in following Chapter 8, has been the most awesome validation relating to the functioning of the human brain. Maharishi had a background in the ancient sciences of the east particularly from the ancient Vedic civilization from where TM originated. The ancient Vedic science unfolded his technique of Transcendental Meditation. Maharishi was a physics graduate from Allahabad University in India. His background in physics and meditation enabled him to base his system of meditation and related it to modern science. Other physicists like Albert Einstein did not have this background of meditation.

REVERSAL OF AGING

Scientific research indicating the reversal of aging through the Transcendental Meditation and TM-Sidhi Programme

PHYSIOLOGY	Normal Aging	With TM Program
Blood Pressure - systolic	↑ 9	↓ 10,14, 98
Blood Pressure - diastolic	↑ 9	↓ 10,14, 98
Cardio Vascular Efficiency	↓ 49, 65	↑ 84
Cerebral Blood Flow	↓ 53	↑ 48
Vital Capacity	↓ 97	↑ 84
Maximum Voluntary Ventilation	↓ 72	↑ 84
Auditory Threshold	↑ 7	↓ 21
Auditory Discrimination	↓ 54	↑ 82
Near Point of Vision	↑ 70	↓ 108
Homeostatic Recovery	↓ 37, 39, 96	↑ 64
EEG Alpha Power	↓ 74	↑ 6, 60
BIOCHEMISTRY		
Serum Cholesterol Concentration	↑ 113	↓ 23
Haemoglobin Comcentration	↓ 27	↑ 48
PSYCHOLOGY		
Suscepability to Stress	↑ 96, 101	↓ 75
Behavioral Rigidity	↑ 89	↓ 78
Learning Ability	↓ 4	↑ 1, 67

All these factors show deterioration with the aging process. The opposite changes – are observed with the Transcendental Meditation and TM-Sidhi program.

Memory (Short Term)	⇩ 3, 90	⇧ 1, 68
Memory (Long Term)	⇩ 109	⇧ 1
Organization of Memory	⇩ 33	⇧ 68
Creativity	⇩ 18, 25, 56	⇧ 61, 95
Abstract Reasoning	⇩ 83	⇧ 78
Intelligence	⇩ 31, 88	⇧ 102
Dichotic Listening	⇩ 47	⇧ 24
MIND-BODY CO-ORDINATION		
Reaction Time	⇧ 112	⇩ 2, 94
Sensory-Motor Performance	⇩ 99, 111	⇧ 15, 85
HEALTH		
Cardiovascular Disease	⇧ 9, 65	⇩ 115
Hypertension	⇧ 17, 26. 38	⇩ 10, 14, 98
Asthma	⇧ 26	⇩ 114
Immune System Efficiency	⇩ 62	⇧ 52
Sleep Disturbance	⇧ 35	⇩ 41
Insomnia and Insufficient Sleep	⇧ 50	⇩ 41
Daytime Sleep	⇧ 81	⇩ 41
Quality of Sleep	⇩ 81	⇧ 41
Depression	⇧ 81	⇩ 34

*A small number of changes resulting from the Transcendental Meditation are not directly opposite to the changes produced by aging. These changes are nevertheless important. For instance cortisol levels and metabolic rate decrease with Transcendental Meditation and with aging. The changes occurring with ageing are related to the accumulated loss of tissue as a result of general wear and tear, whereas the changes occurring with Transcendental Meditation are associated with decrease stress levels and greater efficiency of metabolic process

In the ancient times long, long ago before religions came about around the world, in ancient Europe, Rome, Greece and even Japan, animism was a prevalent religion. Animism was known to be the most ancient religion in mankind's history.

David E. Comings, M.D[22] narrated the origins and evolution of major world religions. He described some inconsistencies in the creationist concept about God created man. Throughout mankind's history man always created the concept of God according to the needs of his time and culture. Religion was formed by men, like it was always in the long forgotten past. Man searched for God in objects of creation like the animism religion or some other forms of worship described in ancient times.

Before religions ever came to be, people worship the sun, the moon, the cow and some other representations of the spirit of God. Some worship animals in India, like the elephant Ganesh. Some worship the trees, others stone monuments.

Animism was a form of worship that relates to the spirit of God present in all things. Animism had a fascination with nature. As history moved on and man evolve further, man sought out man - gods, not anymore nature objects or the sun or animals. The Pharaohs of ancient Egypt and the Emperor rulers of Japan became the man - gods worshipped by the people.

It was said that the Apostles made Jesus Christ, a man-made God. Jesus publicly declared that he was the Son of God. This created controversy over the concept of the Jewish religion that there is only one God. It was a heresy and punishable by death at that time and place, to declare that there was more than one God especially by the prevailing Jewish religion. Jesus Christ was crucified because he declared he was the son of God.

For this reason, early Christian scholars created a trinity dogma to protect the infallibility of God the Father, God the Son and the Holy Spirit. It became a trinity of God. Christians followed the doctrine of the Trinity of God and many of the new Christians of today still follow the same doctrine of trinity.

Religious historians found 100,000 religions in mankind's history. It is because the deity or God particle is part of the human brain. Therefore, throughout history man created religions according to the prevailing culture embraced at that time. The God particle does not evolve. It is inherent in the human genome. This spirituality particle is part of the brain at birth. It is inherent and deeply engrained

---

22 *Did Man Create God?, 2008, David E Comings, MD*

in the human brain. Soon in time, from man's brain as the organ of civilizations, advanced human species evolve with better intelligence and reasoning. The finest and primal particle or the "sap" that Maharishi referred to is hard-wired into the human brain at conception that is why man will always seek God in one form or another. The degree of stress saturation in the human brain cause inconsistencies and aberrations in the interpretation of God that is why war and terror groups break out in the name of religions. TM has been repeatedly validated to remove stress.

Science mirrors the form and power of God in the calming of storms and tornadoes, balancing the climates, countering terror groups and the ending of wars, the improvement of the economies of nations and the fading of all negativities in the world. These are evident manifestations of the power of God within man. This is the Maharishi Effect functioning within the human mind in large groups together at the level of silence!

Science illustrates that if particles are contained together like a tank of hydrogen particles, a huge blast can be created. Similarly, if the bliss particles, the primal or the bits of God particle, are placed together in one place, formed in large groups of human brains, a huge blast of bliss is created blasting away all negativities of life on earth!

Maharishi described that it "creates from itself, by itself and within itself." If so, then it could be the God particle or the particle of everything that scientists refer to that created the universe including man. Maharishi referred to it as the seed form of natural law. This "sap" or solo particle "create from itself, by itself and within itself" is the seed of nature described and foretold the theory of God at the level of the Unified Field. It truly looks like it is the spirit of the anima solo particle I was referring to that is present in our mind, in our body and present in all forms of creation.

It is like the finest elementary particles of nature that animates all things including the stars and the galaxy above. Science may have come up with a concept of the God particle that may have blasted off the creation of the universe at the time of the Big Bang. Maharishi said that the Unified Field of the universe is enlivened and becomes alive in the human mind during the practice of Transcendental Meditation.

With the illustration of the Unified Field on Chapter 8, I was tempted to conclude that the concept of our mind and body as the conductor of the Unified Field energy has an undisputable basis.

Both physicist Maharishi and Hagelin referred to it as the field of pure consciousness in the human mind. Maharishi's background of meditation that he learned from his teacher from the Himalayas was a tool for him to conclude a system of the mind that is in accord with nature.

It is a system to create a problem-free generation. It describes how the finest elementary particles of nature create a force field as illustrated in the behavior and unification of the elementary particle energy in physics shown in Chapter 8. Maharishi postulated that these force fields unite at the most fundamental level as one unified energy field, the root of all energies in nature.

See the illustrations of the unification of the forces of nature in Chapter 8 - Fig. 1 & 2. The Maharishi's Technology of the Unified Field illustrates the unification of all the four fundamental force fields of nature into one single seed form of energy. It is the very root of all energy in nature. The four fundamental forces in nature electro-magnetism, weak force, strong force and gravity ushered in the most significant knowledge benefiting man.

Electro-magnetism has been televised as one system of determining the presence of oil underneath the earth's surface, thus familiarizing the general public with the existence of the electro-magnetic field. Maharishi's correlation was correct our planet earth is surrounded by electro-magnetic field.

Going back in time during the sixteenth century, Galileo was tried for heresy by the Church Inquisition and was placed under house arrest for the rest of his life for proclaiming that the earth was round and not flat. He proposed his theory of heliocentric earth, not flat earth-theory for which he was persecuted by the church.

When the telescope was invented, it enabled astronomers to see the various rounds of planets floating in orbit around the sun in our galaxy and validated that our planet earth was actually round, not flat after all. We gaze at the stormy sky and see the bursts of lightning indicating the presence of electricity in the air.

An object falls to the ground proving the existence of gravity. It was all it took for Newton to discover gravity. A fruit fell down on him while sitting under a tree. Gravity keeps all matter on earth stuck to earth and not fall off the earth's surface.

The falling asteroids must have burned off their graviton particles that is why they fall off their orbits! The compass points to one direction, north, south, east or west. It shows the earth has a magnetic field. Einstein found some force field around the earth's surface. There are atoms and some unseen particles moving and floating around in space. He came up with theoretical physics illustrating the energy of the strong force that binds atoms together.

Einstein's study of the atoms became the fore runner of atomic physics. Einstein elaborated how atoms are bound together by the quantum properties of protons and neutrons. Einstein never liked quantum physics. His work focused more on gravity and its effects on space-time. Some other physicist like Hideki Yukawa's theory of the mesons collaborate the discovery of the strong force.

Many contemporary physicists have brilliantly described the four forces of nature in detail. Some extensively pursued Einstein's quest to find the Unified Field. Some scientist particularly researched on the Grand Unified Theory (GUT) with gravity still in transit.

Gravity frustrated many scientists including Einstein himself who died before he can prove that the Unified Field exists. In the Unified Field theory, Maharishi included gravity as part of the unification of the four fundamental forces of nature. All the four forces of nature unite as one and the laboratory of unification is located in the human mind. It brings man to the level of the ultimate and supreme science.

Michio Kaku[23] a contemporary physicist wrote, "The existence of the universe depends on consciousness. This does not have to be human consciousness but some sort of intelligent life somewhere else in the universe..."

Maharishi emphasized that the human consciousness is the seat of nature in seed form when the human mind is placed at the level of silence. Man is part and parcel of the constitution of universe and his brain can tap the field of intelligence in nature which operates the whole universe.

The concept of both physicists, Kaku and Maharishi appears to meet somehow at some level of intelligence - Kaku's "somewhere else in the universe," and Maharishi's "intelligence of the universe located in the human mind." Lecomte[24], another physicist said that the human mind evolves to an intelligence that is nature's ultimate goal. As another physicist, Stephen Weinberg, who has done extensive work in particle physics, once remarked, "Nature is much simpler than it looks. Nothing makes me more hopeful that our generation of human beings may actually hold the key to the universe..." Maharishi stated that the science that has unfolded in our time, referring to the science of the human mind at the state of silence is the ultimate science because it is the root of all sciences. It is the seed form of natural law.

---

[23] *Physics of the Future, 2010, Michio Kaku*
[24] *Human Destiny, 1962, Lacomte Du Nuoy*

The key unlocking the secrets of the universe depends on the consciousness of man. When the 9,000 is formed, maybe further knowledge about the universe will unfold. This describes the great inner power of man that needs to unfold. The level of silence in the human mind has ushered in the age of enlightenment. "When goodness starts to flourish instead of non-good, enlightenment is at hand," Maharishi said. "The level of silence, is the level of all purity, all goodness and the field where the seed of nature exist." Maharishi said. I concluded it must be where the anima solo particles are located. Only a handful of enlightened individuals are needed to keep the truth going, Maharishi predicted before he passed away.

The Rajapark College in Thailand had some seven hundred students meditating together at one time. There are fluctuating numbers, residents and students meditating together at the Maharishi University of Management in Fairfield, Iowa, USA since more than forty years.

Fairfield has now become the TM town of America where Maharishi Vedic City was incorporated in July 25, 2001. Maharishi's plan was to establish a group of more than 9,000 thousand TM practitioners in our time, in one place, it became the greatest endeavor of Maharishi's life. It has become mine too and wish I had the means to create it again. The number increases as the world population grows.

One, two or few individuals may come to fulfill this dream. It is the greatest endeavor ever. Removing the problems of man and his environment has become the greatest quest of humanity. No one has come up with a permanent solution to solve many human problems occurring daily. Maharishi came up with the Maharishi Effect! It came out to be the most amazing and significant discovery of the twenty-first century in the effective transformation of man and his environment!

Any leader of any country can implement Maharishi's theory of government to produce a problem-free government. One leader, a President, sometime in 1994 implemented the theory and validated the effects. Then in 1994, President Chisano of Mozambique, Africa tried his military to practice the Transcendental Meditation program in groups. Mozambique was at war for more than twenty-five years and their land was dried from drought at the same time. Soon after meditation started done by the military, rain fell on the parched lands. War ceased and significant positivity prevailed in the country. The IMF loaned money to Mozambique that started to improve the country's economy. Mozambique's economy rose 19% when the military started to meditate and became the fastest growing economy in Africa in contrast to being the poorest. New found of coal, natural gas and petroleum gave employment to thousands of its citizens. Just like during the Taste of Utopia in Fairfield in 1984, the stock market rose ten–fold in just three weeks of assembly as shown on Chapter 6, Chart 7 and Chart 6. Also, new discoveries were uncovered around the world as Maharishi has foretold.

Truth and goodness will come grounded on the scientific theory of the quantum field, the Unified Field produced when man quiets down his mind. This level of silence produce the corresponding positivity, a result of Maharishi's programs of TM and Yogic flying.

Quantum physics illustrates the attributes and behavior of the elementary particles in nuclear physics and quantum mechanics. Maharishi has postulated that the quality of the mind at the silent state participates in this quantum field. A water particle has two parts hydrogen and one part oxygen. Hydrogen and oxygen contain a number of electrons and protons similar with other atoms. Earth has water and matter. The stars above has the carbon particle that glows in the dark. Man has liquid water in the blood and matter contents in his bones and muscles. They contain fermions and bosons just like contained in the sky and in the whole universe.

Man is a container of the particle energy field within and is able to harness the energy that animates the whole of the universe. When man quiets his mind thru TM, he becomes an active participant in the functioning of natural law in its seed form. He draws energy from the root energy field that nourishes life. This flat form of silence elevates man to the highest and most superior form of scientific specie. Man participating in the functioning of natural law and thus harness the ultimate, finest and purest energy of life within and outside the whole universe.

8

THE DISCOVERY OF THE UNIFIED FIELD

Ancient philosophers and scholars referred to earth, water, fire, wind and ether as the constitution of the universe. Modern physics expounded the building blocks of matter on earth and in the heavens in terms of molecules, atoms and sub-atomic particles and the defined force fields they create, like the electrons in the electric field, magnetos in the magnetic field and the graviton particles in the gravitational field. The Unified field is a recent find brought to light by the late Maharishi Mahesh Yogi which finally unfolded the capability of the human mind at the level of silence to transform his environment. The discovery of the Unified Field elevated man to participate in the functioning of natural law. Maharishi said, "At the most silent level of the mind, man become master of natural law."

About thc ycar 1600, scholars and the church differ in the concept of the laws of the heavens and the laws of the earth. The laws of the heavens were under the unquestionable territorial authority of the Church and ordained by God. The laws of the earth were something else. Giordano Bruno was burned at stake in Rome when he stated that the sun was just another of the many stars in the heavens. A decade later, a physicist and astronomer at that time, Galileo Gallelei made a statement that the earth was not the center of the universe and it moves around the sun, and that the earth was round, not flat. He was placed under the Inquisition and arrested by the Church. In the following century, the earth was found to move around the sun and is round, not flat after all. The earth is not the center of the universe but part of a multi universe,

In 1687, Isaac Newton united the laws of heaven and earth when he discovered the laws of gravity where very particle of matter in the universe attracts every other particle with a force that is directly proportional to the product of its mass and inversely proportional to the square of the distance between each other.

Newton was able to calculate the motion and the mass of each planet and related it to the universal law of gravitation. Newton then concluded that the laws of the heavens are the same as the laws of the earth, thereby uniting the laws of heaven and earth.

Scientists soon pursued the quest for one single Unified Field, the unification of the four fundamental forces of nature, electro-magnetism, weak force, strong force and gravity.

In 1860, electricity and magnetism was found to be the same face of the coin. Maxwell discovered that when a magnet is shoved along a coil of wire, electricity was produced and vice versa, when an electric current is run thru a coil of wire, magnetism is produced.

Albert Einstein, in the early 1930's extensively sought for the Unified Field and believed that it exists, but he passed away without locating where the Unified Field can be found. Many scientists after him persistently pursued the Unified Field.

In 1945, the unleashing of the tremendous power hidden in the nuclei of the atom brought about the development of the atomic bomb that destroyed the whole city of Hiroshima putting an end to World War II. The power of the hydrogen particle was discovered by breaking down the atom and the group effect contained in form to be a strong force of exponential magnitude that caused the powerful atomic bomb blast.

Similarly, the finest particle of nature found to linger within the human mind, when placed in large groups together create the Unified Field that can put an end to conflicts and all wars. The Unified Field does not destroy, on the contrary, it nourishes all life. In large groups together, at the level of silence, man has the capability to create peace and prosperity for his environment and the world

In 1974, The Grand Unified Theory (GUT) was formulated. The GUT was believed to unify the three forces including the strong force that binds the atoms together. Three physicists, Steven Weinberg, Sheldon Glaslow and Abdus Salam came up with the "electro-weak" unification. These three scientists understood that the weak force can be treated as the same force as electromagnetism.

In 1983, after a global experiment called the Global Taste of Utopia called for by Maharishi described in Chapter 6, the tremendous exponential influence of the Maharishi Effect was confirmed to affect the world. Maharishi confirmed that the fourth force, gravity, can be united at the level of the Unified Field and the point of unification can be found within the human brain at the level of silence. The human mind can handle gravity when the body lifts up at maximum brain wave coherence defying gravity and displayed by the body lifting up during Maharishi's Yogic flying program.

Man now has reached a territory far beyond the heavens mastering the fundamental laws of nature, at the level of silence thru Maharishi's Transcendental Meditation and TM-Sidhi or Yogic flying demonstrating that the human mind can defy gravity when the body lifts up at the level of the Unified Field, at a single point value where all the laws of nature unite.

With the recent and stunning achievement in space technology, man has landed on the moon and established an international station firmly floating in space. For more than fifty years, man explored the laws of the heavens and recently landed a spectacular robotic spacecraft Curiosity rover to Mars. The exclusive dominance of the heavens by the Church long, long ago is gone. Finally, science took foot hold of the heavens that was once controlled by religion. The heavens have now become the playground of particles and a new science has begun to unfold the infinite property of space that holds all the planets of the universe. Space once thought to be empty is not empty after all, it is filled with nature's fine particles vibrating in space.

The grip of the old church began to decline as more scientist explored space and the heavens above the earth. Science began to loosen the old norms of religious dominance. Stephen Hawking, another well-known contemporary physicist stated that "the creation of the universe was made by the laws of physics and doesn't need to be set in motion by some god." [25]

It was the theory that the elementary particles of physics united through the interaction of energy and matter fields resulting from the super particles that could have produced the Big Bang, the theory that described the beginning of the universe.

Maharishi's Unified Field Chart illustrated in Fig. 1, in the following page, show how the unification of energy and matter fields spring from the Unified Field at the seed of creation.

If creation started with a Big Bang as scientists postulated it to be the origin of the universe, it could have come from a concentrated form of expanding process that blasted off these finest, unseen anima solo particles, the AS_U particles that animates and nourishes the whole universe, that Maharishi said it created "from itself, by itself and within itself" as he described the unseen, non-material, indivisible "sap" of the universe. His theory, as a physicist stipulated that the "sap," these finest particles can be drawn by the human mind through the practice of his Transcendental Meditation technique of quieting the mind. Maharishi called this technology of the finest particles and the field where all the forces of nature unite at the seed form of nature, the Maharishi Technology of the Unified Field, in the field of consciousness.

The anima solo, the solo, the ASu, the finest particle at the root of creation is the life-giving and animating finest particle of

[25] *Grand Design, 2010, Stephen Hawking*

nature. It produces the purest energy of nature. It is a sole particle by itself. It must be million and million times remotely finer than the tiniest atom discovered to date. Scientists say, the finer they become the greater power they produce that is more invincible from outside influences than the magnetic or the gravitational field since it is the finest seed form of natural law. Atoms are measured by megatons like displayed by the yields of force as nuclear or atomic blasts. The anima solo particles are not measured by tons, but their observable effects can be displayed by the number of human minds placed together at the level of silence. The Maharishi Effect display and measure the coherence and spectacular positive trends it produces in society referred to in Chapter 6, in the Global Taste of Utopia which unfolded the magnitude of positive trends to prevail on earth. Maharishi postulated that the result has a nourishing effect that is manifested and produced by the billions of human brain cells functioning together at zero fluctuation of the human mind during Transcendental Meditation and Yogic Flying.

Maharishi has said that when one percent of the population practice TM or a smaller number equal to the square root of one percent of the population of the world or any country, who practice the advance TM-Sidhi program or the Yogic flying technique that Maharishi introduced, in one place together, all life will be in the positive evolutionary direction of natural law. All positivity will prevail. No negativity can penetrate the invincible Unified Field. (See demonstration of flying next page). At the maximum coherence of the human brain describes the junction point of all the forces of nature, they unite at the level of the Unified Field, then the body lifts up in Yogic flying. The three fundamental forces of nature, electromagnetism, weak force and strong force unite with gravity at the level of the Unified Field within the human mind illustrating that gravity can be handed by the human mind when the body lifts up defying gravity. Since the Unified Field contains the finest and purest of all particles of nature, it has the character of invincibility when practiced in large groups together in one place. No negativity can exist within the silent level of the human mind in large groups together.

The anima solo particles can be drawn by the human mind and the nervous system act as the conductor of the Unified Field. It is like the copper wire that acts as the conductor of the electrical field in the atmosphere. The human brain and nervous system become the instrument, conduit and channel of natural law. Maharishi said that

Maharishi's TM - Sidhi program, the Yogic flying technique illustrates the mechanics of world peace.

this "sap," this particle holds the total potential of nature. It is the intelligence of nature within itself. It contains totality of natural law. It is an eternally formless silent particle. It is the purest and most powerful of all the particles in nature. Maharishi said, "It is the knower, it is the process of knowing and the known" value of nature. It is like the creator, the process of creating and that which is created, all three in one, like a trinity.

The unification chart came from Maharishi's Technology of the Unified Field further illustrated in the following Physics Chart FIG. 4 in the following pages. I further illustrated the unification of the energies of nature rooted like a tree. Since it is the seed form of natural law, I illustrated it as a seed and shown at the root. The root energy is the seed form of nature where all energy of nature spring from. Gravity springs from the root. The strong force binds atoms together, weak force, electricity and magnetism unite as one; and all the other forces of nature are rooted at the base, at the Unified Field. The human mind becomes the ground and basis where the seed of natural law grow. Man has now the capacity to handle nature from its seed! Man become "master of natural law," just as Maharishi has repeated many times.

Maharishi illustrated the unification of the four fundamental forces of nature in Fig. 1. I illustrated it as a tree in Fig. 2 and Fig. 3

as shown in the following diagram. Many scientists besides Albert Einstein, Stephen Hawking and Michio Kaku have extensively described the same four fundamental forces of nature that Maharishi referred to.

As the Fig. 1 shows, all the forces of nature are unified at the level of super-unification. The conclusion is that the Maharishi Technology of the Unified Field, being the technology of the finest, most basic level of natural law, it is superior and more powerful than electronic, nuclear, and all other technologies at any level of nature's activities. Another conclusion is that the future of all scientific research lies in the Unified Field and the future progress of all mankind in the Maharishi's Technology of the Unified Field, which is the technology of the subjective aspect of life and is be the basis of all objectivity in nature. Like the objective tree, that is manifested from the ground, the subjective root, that part of the tree that is the hidden from view, absorbs the "sap" from the earth and makes it grow. Similarly, the "sap" hidden within the human mind, makes man grow to fulfillment. Stress blocks the growth of man to fulfillment and prosperity.

The subjective aspect of life, the root of all creation was foretold by masters from the ancient Vedic civilization. The ancient Vedic masters must have preserved the ancient texts held by man from birth and that which kept man happy and content since the origin of human civilization. In the ancient Vedic civilization, the ancient Vedic literatures were chanted daily. Its texts and verses contain the fluctuation of natural law and are the expressions of the human physiology.[26] Poverty did not exist then. They draw that silent "sap" of life, the Unified Field, giving sufficient satisfaction to all their needs. In the modern world, people are more drawn to the objective value of life. They glorify flashy cars, luxurious homes, travels, clothes, jewels and other material pursuits that satisfy personal egos. The life of luxury appears to be every ones dream. In the long history of man, no one seem to be interested with the inside subjective aspect of life. It was left for the ascetics seeking solitude and lived in bare necessities. Even then, asceticism gradually faded away as it does not fully satisfy the true nature of man.

Maharishi's said that the birth right of man is two hundred percent of life, one hundred percent inner happiness and one hundred

---

[26] *Veda and Vedic literatures, Expressions of the Human Physiology authored by Tony Nader, MD, PhD, First Ruler of the Global Country of World Peace*

percent outer fulfillment. The Unified Field holds the total potential of natural law. It is the field of bliss. At the level of bliss, desires are fulfilled. It satisfies the goal of man gaining one hundred percent inner happiness and one hundred percent outer affluence. Like a tree, nature so provided the "sap" at the root that completely makes all the aspect of the tree to grow, the leaves to turn green, the flowers turn red, yellow or white and consequently bears fruits. The tree naturally blossom in its full glory. It is the nature of man to blossom in its full glory too, with a brilliant mind, healthy body, wholesome behavior and live progressive and at peace. The root of man was missed for many centuries that is why man started to suffer. The Unified Field completes human life, inner happiness and outer affluence, both hand in hand completes the true nature of man. The unmanifest level of human life can be experienced at the non-material, unseen level, at the Unified Field, at the root. It is unseen and unmanifest, but the effects are seen. Like electricity, we do not see the electrons and protons. We only see the effect, the light that is produced by the electrons. When man draw at the unmanifest field, the Unified Field, it is the field of bliss and man experiences peace. The basis of peace is bliss. Man can tap that field of bliss within his consciousness and enjoy full glory of life. When the particles in the electric field become heated from the sun, the hot particles collide with the cold air particles in the atmosphere, producing lightning which confirms the presence of the electricity in the air.

It displays the non-material, unmanifest subjective field of nature, the field of unseen particles. Like air, it is unseen, but when cooled, then we see the rain, air in liquid form. All these fields are unseen by the human eyes. They are unmanifest yet they truly exist. The root energy field is available in the human brain when the mind is at a state of silence beyond thought fluctuations. Thoughts disturb the phase condition of the energy field which is so fine that it can only be drawn at zero activity of the human mind. Transcendental Meditation quiets the mind.

The human skull protects the brain from outside disturbances, yet the faintest thoughts disturb the finest fluctuation at the level of the Unified Field, the root energy field lingering within the human mind. That is why the mind must be placed at the level of silence. The finest and purest of all energy of nature can be drawn at the quietest and finest level of mental state. Like the electrons and protons in the electrical field and gravitons in magnetic field, they have no appearance to the naked eye because they are invisible, yet these

THE FOUR FUNDAMENTAL FORCES OF NATURE

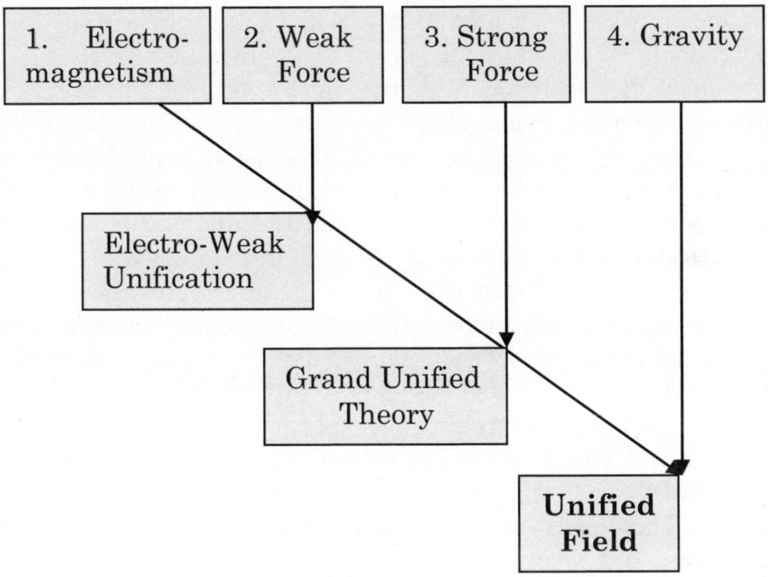

FIG. 1 MAHARISHI'S TECHNOLOGY OF THE UNIFIED FIELD

particles exist and create force fields everywhere in the atmosphere around the earth and across the expanding universe.

Fig. 1 illustrates the historical unification of all energies - forces in nature, electro-magnetism, weak force, strong force and gravity unite at the root, at the Unified Field. Maharishi described the Unified Field, as the "sap" of the universe, the root energy field of nature, like the "sap" inside a tree. The "sap" permeates from the roots of the tree up to its leaves and its branches. It is the energy that makes it grow.

The Unified Field is the seed form of natural law that creates all the diversity in nature. Because these particles are so fine and pure, it can be drawn by the human brain only at the level of silence. These energy particles are the tiniest beyond measure. Their manifestation can be measured only at the present time like the effects of the group practice of Transcendental Meditation and Yogic flying.

Scientists described the concept of the Big Bang which was projected to have occurred about some 15 billion years ago and was presumed to be the average age of the universe. Scientists postulated that the four fundamental forces of nature were assumed to be originally united at the point of the Big Bang. They must have begun as one singular and primal force. At the initial moment of the Big Bang, scientists calculated that gravity first separated at 10^{-43} seconds after the first blast from the condensed high energy particles, like a tiny blast of energy described as the point of the beginning of the universe.

These are just assumptions and calculations made by scientists. But what actually happened at the exact beginning of creation, no one knows. I still think that there is a God, the perfect designer who organized the orderly functioning of the universe. Until then science may one day find how the design of creation came to be.

The weak force was the last to separate. The weak force resulted from the radioactive decay produced by the spontaneous collision of particles. Scientists approximated that the strong force separated at 10^{-34} and the electromagnetic force at 10^{-12} seconds when the universe expanded many billions of years ago. The universe is still expanding infinitely.

No one exactly knows where and when the high energy particles came from that blasted off the universe from the Big Bang. The theory stipulated that it created from "within itself, by itself and from itself," like Maharishi's "sap," created the theory of the Unified Field where all the four forces of nature spring from. The "sap" of the universe at the level of the Unified Field created "from itself,

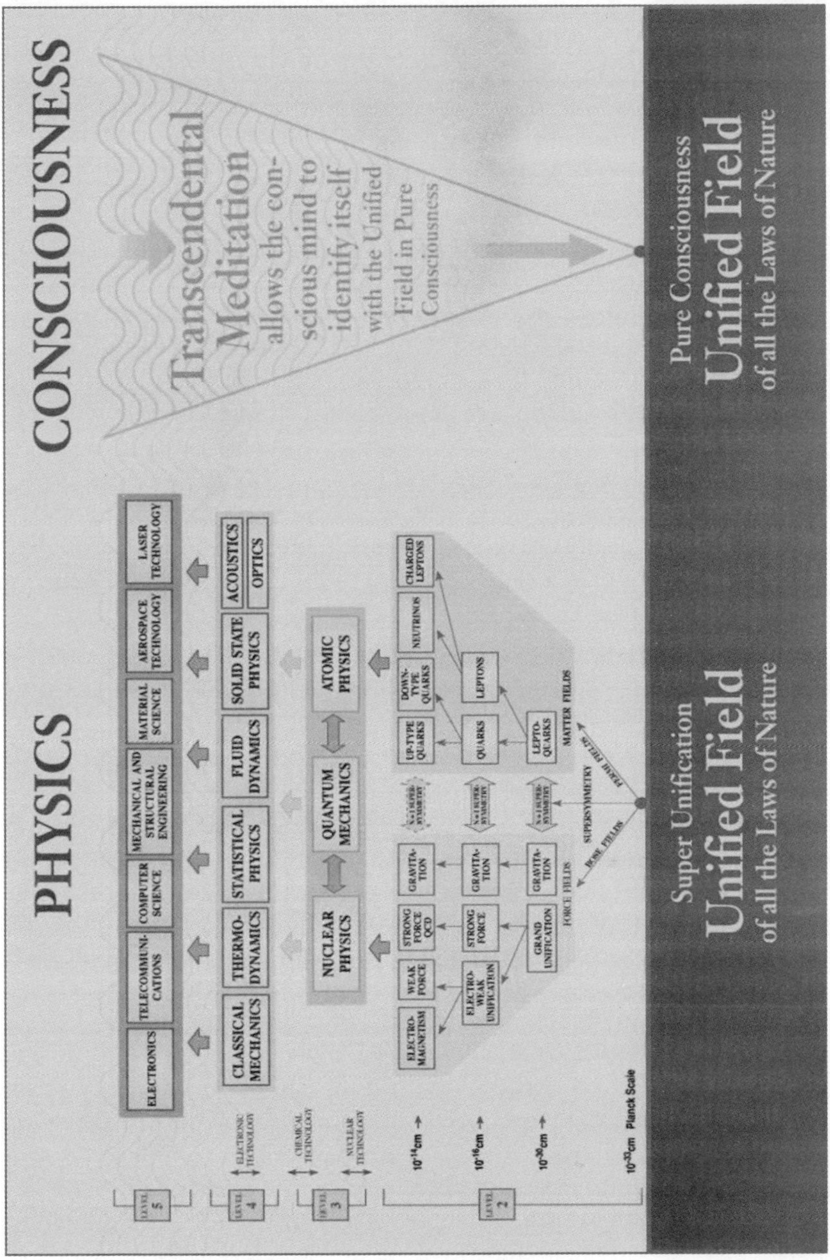

FIG. 4 MAHARISHI'S PHYSICS CHART[27]

[27] *Global Country of World Peace*

within itself, by itself," Maharishi said. "It is the seed form of natural law." What is important to know is the Unified Field particles linger within the human mind at the level of silence during the practice of Maharishi's Transcendental Meditation program. This elevates man to the level of the ultimate and supreme science and the anima solo particle that lingers within the human mind must have participated in the creation of the universe. That part of the mind in creation is a baffling theory that many scientists today have yet to explore.

Some scientists say that a system does not exist unless it is observed. It means that the mind is a critical part of the universe. The observer is significant in the result of an observation and the observed. Maharishi repeated the same concept of the "Knower, the process of knowing and the known or the observer, the process of observation and the observed." Like a trinity existing within the human mind. But until then, no one knows how the universe came to be, someday soon, science may confirm the speculations that the mind partook in the creation of the universe. Meantime, let a layman believe that there is a God that created the universe. I too believe in God. This strange concept of trinity kept me in awe and wondered if it was the anima solo particle that has the attribute of the total potential of nature and posses the intelligence of nature in seed form, like three seeds in one fruit! Like many seeds inside one pomegranate fruit, like an apple with five seeds, or like one flower that has four petals instead of five. Some flowers have five petals. The rose may have more, but all have just one stem from its roots unseen from the ground.

As Einstein had said, "God does not play dice." It must have been a bit of God's own particle left in place within the human brain so that man can easier explore the laws of nature than exploring the theory of the Big Bang or the beginning and origin of the universe.

The spectacular documented data of the Maharishi Effect at the Global Taste of Utopia event on Chapter 6 displays the grace and the will of God on earth. The magnetic field envelopes our planet earth and protects it from solar heat and solar storms, otherwise all existence will not be able to survive the solar heat from the sun.

The voltmeter measures electricity. The tiniest particle of nature manifests itself through the effects of the Unified Field within The human mind at the level of silence during TM. These tiny particles find resistance to the slightest form of disturbance at the slightest thought fluctuation of the mind.

The Unified Field becomes an invincible field, free from any outer negative influence. It contains all positivity in nature where all diversity in nature are created from its seed, like a tree creating the leaves and flowers at its root. The Unified Field is the root of all creation. The root energy particles are the finest form of energy, beyond electrons, protons, gravitons, bosons, fermions and beyond other fine particles discovered to date. These particles are subatomic, which means, they are smaller and finer than any known atomic particles. Scientists attempted to measure fine particles by installing a particle accelerator five hundred feet below the earth's surface at the depth beyond any disturbances and any fluctuations. Faintest fluctuations of fermions were measured at this deep level from the earth's surface.

The presence of the field of gravity on earth keeps all earth matter particles including man from falling away from the earth's surface. All matter on earth stick to earth as it moves along with all the other planets because of the force of gravity.

The weak force relates to solar energy and the radioactive decay of atoms. As an ordinary plain layman observer, one sees a sunflower harness the energy of the sun which blossom from the east when the sun rises until full bloom and close its petals when the sun sets in the west. It is apparent that the sunflower harnesses the weak force from the sun. The morning glory flower blooms precisely at the same time in the morning as the sun rises. Man wakes up upon the rising of the sun and man start to feel sleepy when the sun has set in the evening. The presence and distance of the sun from the earth affects the functioning of the human mind and body. This illustrates to a layman how the energy and the distance from the sun affect man and life on earth.

Maharishi said that the silence within the human mind connects man with the Unified Field of nature. The Unified Field allows man to live life in accord with natural law that is problem-free. He said this level is the most fundamental level of nature in its finest form that is why the mind has to be in the finest and most silent state. Transcendental Meditation can easily quiet the mind in an effortless manner and contact the flat form of natural law in its seed form. Natural law is the law of perfect order. At zero excitation of the mind, these finest of all particles at the root energy field provides nourishing and animating energy. It is available for the human mind to draw from in silence. It is the total potential of natural law and allows man to live in accord with nature that is problem free. The root energy field provides perfection in man's life beyond negativity. Because the root energy, the Unified Field is beyond negativity, just as in the electric field, no impurity can exist. It is a natural phenomenon in nature. Maharishi calls the Unified Field the most superior of all technologies because it refers to man's own well-being and it is the most supreme knowledge connecting man to nature and the universe. It captures the birth right of man to live life without problems.

The Unified Field, the root energy field of nature defines the relationship of man and the universe. It is the energy that administers the orderly functioning of nature. It defines the same order and harmony within the human physiology at the level of silence.

So too can man harness the orderly functioning of nature within the human mind and nervous system in the most nourishing manner that is perfect, beyond negativity. It is because nature is perfect and thus allows man to live in perfection in accord with nature beyond problems and suffering. Man is part and parcel of the constitution of the universe. By nature, no negativity exists at the level of the Unified Field. It is nature in its purest form.

In similar manner, no impurity can exist in the magnetic field, like when you place iron filings on a piece of paper with impurities with it, and place a magnet beneath the paper, the impurities in the iron filings move out of the magnetic field. No impurities also exist in the magnetic field, by nature.

Maharishi has said that the future progress of mankind lies in harnessing the Unified Field which is the supreme and most powerful energy than electronic or nuclear technologies. The laws of nature are far-reaching beyond human made laws. Natural law dictates the result of every action. Boil water at one hundred degrees and the water particles turn into gas in the air and it is made of two parts hydrogen and one part oxygen. The liquid water becomes the air we breathe as it changes into air particles when water evaporates thru the application of heat.

Likewise reduce its temperature to zero degrees and the liquid water will solidify as ice crystals. By nature, no impurity can exist within the ice crystals at zero degrees. By nature also, at zero excitation of the mind, no negativity should exist in the life of man. At zero excitation of the mind, no negativity can occur because it is the seat of the purest and indivisible particle of nature. This is the basis of natural law in the human mind. When the human mind is turbulent, the surroundings become likewise turbulent and negativities prevail.

When nothing else can solve the problems of man thru man-made governments, try the Maharishi Effect. Introduce the science of the Maharishi technology of the Unified Field and validate the results.

Introduce the system of Transcendental Meditation in government and expect the positive results thru natural law administration.

It is the greatest endeavor for man to live problem-free. Nature's government within man is more orderly and more superior than any outside man-made governments. Nature provided man the attributes of infinite creativity, bliss, field of all possibilities, invincibility, balance, freedom, nourishing and total potential of natural law. These are the character and attribute of the Unified Field. It functions in simplicity and in silence in the vacuum quantum state. It is the spring board of all orderliness in all of the diversities in nature. In man, with diverse religions, diverse forms of governments and diverse races, each and all men belong to one planet earth drawing nourishment from the same "sap" and from the same unified root energy field. From zero to diversity, from zero to many, all spring from the Unified Field. It creates from within itself. Maharishi said "Rishi, Devata and Chandas," referring to the attributes of the mind in Sanskrit language. "It is the seed formula of creation. The seed is within itself in the vacuum state, in silence, creating all the diversity in nature.

"See this rose," Maharishi holding a stem of a rose flower, "the sap inside tells the leaves to turn green, tells the thorn to be a thorn or to take the color red, yellow, or white rose petal. There is order by design in all the diversity in nature." Maharishi's many long discourses about natural law says that from the "sap" inside it spreads to the diversity of the leaves, the thorns and the flowers. The "sap" creates within itself in silence. The same "sap" enlivened within the human body creates the bones, the hair and everything within man.

The "sap" of nature, the finest particle in the Unified Field, creates within itself the billion cells creating the whole of man, white, brown or black. Like a rose flower, it comes in red, yellow or white. Man spring from the same "sap."

The sun, the moon and all the planets spring from the same sap creating the whole of the universe. We can tap that spring board of nature when we quiet our mind through the Transcendental Meditation technique. Man becomes an active participant in the functioning of nature at the level of silence.

Quantum physics describes how particle energy fields become more powerful at the vacuum state. The nothingness or emptiness contains the most powerful particle energy field in nature that is far beyond nuclear and atomic fields.

It is important to understand conductivity in nature. Like the copper wire, it is the conductor of the electrical field in the atmosphere. Similarly, the human brain and nervous system is the conductor of the root energy field, in the Unified Field that is present in the atmosphere.

Now, it becomes possible to harness this field and contact the attributes of the field, which animates the silent and perfect movement of the planets in the universe, including microbes, plants, animals and man. These tiny particles are in the air, the same as the protons and neutrons in the electrical field also present in the atmosphere. Without this root energy field, all things die. The stars collapse and fall. Just like when the copper wire is fatigued or corroded, there is resistance in the flow of electrical current within in the copper wire. Similarly, if the mind is turbulent and the body is fatigued and full of stress, there is resistance to harnessing the attributes of the finest form of root energy field, the Unified Field.

When the mind quiets down, the root energy particles restore balance in the body. It is by nature the attribute of these root energy particles to nourish, just as by nature, water nourishes plants, creatures and man. When the mind and nervous system is continually habituated in that supreme level and state of consciousness, at the level of the Unified Field, there comes a time when all stresses of mind and body disappear.

At this state, Maharishi said, "I am Brahm, I participate in the infinite unity of the universe. I am totality. My awareness floats into the intelligence and awareness of nature. The mind attains its total potential in silence. Man becomes master of natural law."

I am awed by Maharishi's statement of becoming master of natural law. It is a statement loaded with deep knowledge of the inner functioning of nature. It assured me more when another physicist Lecomte Du Nouy that the human brain is the masterpiece of nature. He said the brain is the main tool in human evolution towards becoming the perfect man.

John Hagelin, PhD and a renowned contemporary quantum physicist who worked closely with Maharishi also referred to the human brain as the field of unity in nature. He described how the human brain physiology can experience and enliven the Unified Field of natural law to produce peace in the world.

I feel assured that these physicists, Maharishi, Lecomte and John Hagelin, repeated the same concept of the mind of man. It became more clear to me what Maharishi had said that we become master of natural law and that when man quiets his mind, man will live in accord with nature.

I concluded that the human brain and nervous system acts as a conductor of the Unified Field of natural law. It is like the copper wire, it conducts electricity. Like a magnet, it conducts magnetic field. This analogy was sufficient enough for me to conclude that the human nervous system is the conductor of the Unified field and can possess its attributes. Like metals that conduct heat energy, it can possess the attributes of the photon particle that transmits the heat.

With all the scientific correlations of energy fields, how else can I conclude that the brain and nervous system is the conductor of the Unified Field energy when man is the beneficiary of the Unified Field based Maharishi's theory? Man unites with the spirit of the universe in anima soma or unites with the source of life in anima yoga, or whatever the ancient seers have called it for thousands of years. It will be the same banana, ancient, past, present-and whatever name it will be in the far future, it will be the same root energy field of natural law.

Science has and will unfold the answer to the mystery of man's happiness and fulfillment. Silence in the mind is one giant step towards the answer. The transcendental state differs from the states of waking, sleeping, and dream state. As the mind attains zero activity and habituates itself in that silent transcendental state, the mind acquires full quiet awareness and body functioning starts to follow that quiet state. Breathing becomes finer, the heart beats less, basal body temperature goes down. Body attains deep rest. Body muscles and arteries relax making blood circulate better. Blood pressure goes down. Imbalances in mind and body start to normalize.

When there is pressure in the mind and body, heart arteries constrict, which causes less blood circulation to the heart, then problems arises such as heart attacks. During meditation, the mind quiets down and the body becomes relaxed, heart arteries relax, therefore heart attacks can easily be prevented with TM. When the body relaxes, blood arteries relax and blood circulation becomes normal. Any physiological and neurological disorder starts to become balanced. Hundreds of researches published in scientific journals have been conducted which confirms the individual health benefits of Transcendental Meditation.

The energy field that animates the whole universe in an orderly manner seeps into the silent mind and the body begins to function in the same quiet and orderly manner. The heart beats automatically. The lungs expand and constricts naturally. The heart and lungs function in an autonomic manner. So does digestion, metabolism, blood circulation and all the various autonomic body functions. The body functions within its own small universe animated by the Unified Field that nourishes life.

The same "sap" of the Unified Field, the energy that animates the functioning and movements of all the planets in an orderly manner likewise provides the same order in the human physiology at the level of the silent mind. Maharishi has said that there should be no reason why body functioning should be in disorder at the level of a silent mind.

By nature, at the level of silence in the human mind, the body should always be functioning in perfect order. To maximize the conductivity of our nervous system and enable us to harness the benefits and attributes of the root energy field, the mind must be placed in silence for the body to gain deep rest and remove fatigue that has accumulated in the human physiology.

TM provides the rest and relaxation that the body needs to regain balance lost with stress that has accumulated daily in the body. Rest is the basis of activity and energy. If the body is tired, conductivity is reduced. Like the electrical current in the copper wire, if there is fatigue in the copper wire there is less conductivity to the electrical current. The body must be free of toxins. Toxins cause disease and imbalance. The body must be nourished with sufficient nutrients to promote growth and replacement of new body cells. Life-style affects health. Maharishi said, "The modern man and the busy executives, always working, running here and there, like rabbits."

"See the elephant, slow but leave deep foot prints with every step. Pull the arrow back one hundred percent and release it, the arrow will fly with great velocity to hit the target. So does the mind. Go down to the deepest quietest level and tap that infinite field of creativity and intelligence within. Like the arrow, just pull the arrow, easily release and it hits its target."

The energy is always available in silence from within. From silence go back to activity and hit the target. Maharishi said, "Do less and accomplish more with my system of meditation." The procedure is easy and effortless. Modern times have the economic and technological breakthroughs to validate Maharishi's Vedic sciences. The Vedic approach to health expands the understanding of the underlying Unified Field of natural law. Millions have learned Transcendental Meditation technique and hundreds of scientific experiments have been conducted worldwide.

After nearly fifty years of Maharishi's history, there are more than enough evidence to show that TM works. There is no reason now why so many people are still suffering. There is no reason why wars still occur, diseases and negativities such as earthquakes and typhoons that frequently devastate the environment can altogether be eliminated by simply forming large groups of TM and Yogic flying together in one place on earth or smaller groups in each country.

With the technology of the Unified Field practiced in large groups together in one place, it is possible to eliminate all problems in one single stroke. Utilize the underlying knowledge of natural law which is the simple natural phenomenon to create heaven on earth now is the greatest endeavor Maharishi wanted done in this generation.

What science wants is a theory that is testable. TM and the advance TM-Sidhi programs have been tested several times and the effects were documented. Maharishi invited governments and scientist to test and validate the results. Like Galileo proved that the earth was not flat and not the center of the universe and that the earth is round. Ptolemy's theory that the earth was the center of the universe and that the earth was flat was corrected 1,500 years after Galileo came to dispute Ptolemy's theory. The Church believed Ptolemy's theory and arrested Galileo for heresy. If Galileo did not challenge Ptolemy, the earth will still be flat.

Maharishi challenged the world without suffering otherwise there will be suffering forever on earth. How pathetic! Maharishi said that our modern times have the means and capability to validate his theory and that many doctors and scientists have done several validations to date.

Galileo corrected Ptolemy. Maharishi surpassed Einstein in locating the finest indivisible particle of nature benefiting man within the brain of man. Science in time always proves the observable truth. The cloak of science governs most of man's functioning on earth. The air he breathes, the food and drug he digest interacts with all the particles of his body. His mind, in pure consciousness identifies itself with the Unified Field of natural law.

Modern award-winning physicist Dr. John Hagelin , one of the great visionary thinkers of our modern times joined Maharishi and confirmed the existence of the Unified Field drawing from the most advanced, mind-expanding discoveries of modern science to the timeless wisdom contained within the most ancient records of mankind, the Vedic literatures, the Unified Field offers a profound, all-encompassing vision of the ultimate reality of life.

Dr. Hagelin conducted pioneering research at CERN (the European Center for Particle Physics) and the Stanford Linear

Accelerator Center. Together with Maharishi participated in the development of a highly successful Grand Unified Field Theory that transformed man to the level of a cosmic participant in the inner workings of nature occurring within the silent level of the human mind during the practice of Maharishi's Transcendental Meditation. Science truly guides the life of man. Transcendental Meditation makes man live in accord with nature. It brings about the ultimate and supreme science of happiness and fulfillment.

The discovery of the Unified Field enlivens its holistic, evolutionary qualities in all aspects of life—physiological, psychological, and sociological. Mind, body, behavior, and society improve. These has been documented by a large body of scientific research showing results such as increased intelligence and creativity, improved health, reduced medical costs, decreased anxiety, increased self-actualization, better job performance and many more traits improve at the individual level. When Unified Field is placed in group form, it creates the Maharishi Effect which transforms the whole surroundings, including climate change. This newest discovery of modern science challenges all traditional forms of individual and global transformation of man on this planet earth.

The creation of a new civilization of heaven on earth is a challenge sought for by man of peace and governments should consider implementing the natural law based system of administration. The human mind now has the ability to create happiness and fulfillment after all the many generations of problems and suffering.

The Unified Field has been sought by many scientists thru many decades. Maharishi found it to occur in human consciousness and can be easily experienced by the human mind during the practice of Transcendental Meditation. Maharishi's discovery of the Unified Field has been the most significant discovery of science to date as it brings about the very root of life and creation in its most silent, unmanifest, non-material form to all diversities in the universe. It nourishes and balances all life.

For man to be the conductor of the Unified Field, the most supreme level of knowledge ever discovered in our generation, the formula for peace, prosperity and happiness for all of mankind is at hand. The goal is to create a group of 9,000 the number needed for the world or square root of one percent of each country's population to practice TM and Yogic flying, then the real purpose of natural and the goal of life will be fulfilled. Climates will balance. Wars and conflicts will end and all negativity prevailing in the world will fade away. The discovery of the Unified Field brings to light the science of creating heaven on earth. It truly is the most significant discovery that will reverse the negative trends seen in the world today. Maharishi has brought to light the root and basis of all life, the Unified Field of natural law. The invincibility principle of natural law is contained in the Unified Field.

The discovery of the Unified Field by Maharishi Mahesh Yogi has brought to light the most significant law of physics discovered in the long history of mankind. The law of physics contained in the Unified Field holds the cooling mechanism naturally possessed by the finest elementary particles of nature prevailing in the whole universe. This law of physics can solve climate change. Scientific evidences already exist which proved that global warming can be solved by Maharishi's discovery of the Unified Field.

This law of nature expressed by the discovery of the Unified Field describes the character and properties of the unseen, unmanifest particles prevailing in space and in the universe. The space previously thought to be empty is not empty after all. The elementary particles of nature vibrate and fill up the space above, below and around each planet in the universe, just like the graviton particles prevailing around each planet creating the gravitational field that keep all planets in place without falling out of their orbits as they move around the sun. The Unified Field posses the balancing quality and mechanism of natural law applicable to climate change and utilizes the finest ability of the human nervous system to sense and react to warm or cold temperatures. This law of physics operates and observed to manifest at the finest level of the human mind in large groups together at the level of silence when the Unified Field principle is applied.

9

EXPERIMENTAL MODELS FOR CLIMATE CHANGE

With the discovery of the Unified Field, experimental models for climate change can be created in a country frequently visited by typhoons like in the Philippines or a state where droughts and wild fires constantly occur like in California. The Philippines has nearly more than two dozen of typhoons occurring regularly each year. In California droughts are constant and many massive wild fires prevailing that raze thousands of acres every year.

Maharishi validated the Maharishi Effect twice in the Philippines and displayed its great effect on climate change. I applied the Unified Field format in the country more than twice that clearly indicated the Maharishi Effect has the same spectacular cooling effect on global warming. It also results in the automatic rise of economic trends and prevailing peace that follows occurring in the environment.

In 1979, a strong typhoon was predicted to touch down in Manila on a weekend. At that time Maharishi was organizing a group of TM and Yogic flying groups at the Antipolo province about two hours drive from downtown Manila. During the week, the participants were manufacturing rattan furniture, at the same time practicing the TM and Yogic flying program before work in the morning and after work. They were called back to work that weekend to avert the typhoon. As predicted, the expected typhoon made a turn-around direction away from Manila to the surprise of the Philippines weather bureau monitoring the typhoon's strong path.

The same occurrence happened in 1986 in Fairfield, Iowa, USA, where groups practice TM and Yogic flying twice daily at the Golden Domes located at the campus of the Maharishi University of Management. The hurricane took a similar turn-around direction away from the predicted touch-down target.

In 1984, Maharishi sent 1,200 Yogic flying participants to calm down the turbulence caused by the multitudes protesting against the Marcos regime at that time. A wild typhoon blew away a ferry boat carrying 1,000 people who all perished at sea. A volcano was erupting at the same year. As soon as the 1,200 TM participants started Maharishi's TM programs, typhoons calmed downed, the volcano erupting simmered away. Under the instruction of then US President Ronald Reagan, Marcos and his wife, Imelda were transported by a helicopter to Hawaii where Marcos lived in peace until he passed away three years later due to an illness. The 1,200 TM and Yogic flying participants created a peaceful transition in the Philippines. A new government was established peacefully without a single shot fired and no victims recorded from both sides, from the protesters and from the Philippine government forces, unlike other government transitions like in Egypt, Libya or Ukraine where many people were killed. In Libya, Maommar Kaddafi was dragged in the streets by the mob to die.

I learned TM in December 1984 and first met Maharishi in Washington, D.C in January, 1985. After that meeting, I had four hundred students meditate together twice daily, in the morning before school and in the afternoon after classes. It was summer that July of 1986, when all the students started their meditation, instantly rain fell and cooled the hot summer climate. There were no clouds foretelling the rain. As soon as the students opened their eyes after meditation, the strange down pour stopped. The TM group balanced the hot summer climate that was prevailing and cooled it down displayed by the instant rain that occurred.

Strange, the drain pipes that were clogged after decades of neglect along the flood zones around the city were instructed cleared by the city's government. So when the rainy season followed after summer, flooding that constantly flowed to the nearby streets ceased to rise. Property values rose significantly as the economy flourished and peace prevailed after the new government was hurriedly put in place.

Exactly the same rain phenomenon happened in another school that I had built in 1994 at a remote poor town of Bolinao, a province located five hours north of Manila that was ridden with rebels in the town at that time. The same strange rain phenomenon instantly poured as soon as the students closed their eyes in meditation, then the rain stopped after the student opened their eyes. There were no clouds foretelling the rain, like in the previous school, it rained and similarly cooled the summer heat that was similarly prevailing that same July in 1994. Even in the rebel held territories in far out southern Muslim Philippines, the rebels laid down their arms, it was all written in the daily newspapers. The rebels in Bolinao town vanished at the same time. Economy rose in Bolinao town and peace prevailed throughout the country. There were signs of economic index that rose in town. An old bridge that was abandoned for fifteen years was rebuilt and the old gravel-based roads in town were concrete-paved. A new two-storey market building was constructed, instead of the open street stores where the town's people sold their wares along the open side streets. The remote poor town of Bolinao instantly transformed with just 500 students practicing the Transcendental Meditation program.

Science proves a theory correct when events repeatedly display the same documented results. The events during the Global Taste of Utopia[28] in 1984 described in Chapter 6, documented the rise of stock markets around the globe and general peace in the world prevailed. Maharishi promised rain would come in drought stricken Brazil, and the long awaited rain did come.

The same rain phenomenon happened in Mozambique in 1994, rain came to the drought- stricken country. Peace prevailed and economic index rose to a remarkable and unprecedented 19% compared to Mozambique being one of the poorest nations in Africa. These transformations happened when Transcendental Meditation and the Yogic flying program was practiced by large groups performed together by the military in Mozambique, Africa.

The science behind climate change is the phenomenon described by the behavior of finest elementary particles of nature,[29] that creates the invincible field around the earth, like the magneto particles in the earth's magnetic field extending from the north pole to the south pole produce the magnetic field of the earth that is invincible from solar flares, otherwise no life on earth can exist. All the planets move harmoniously in space around the sun without falling out of their orbits. They are held by the all powerful graviton particles in the planet's gravitational field.

The nuclei of the atoms, when they break down to finer particles, they create a field. The finer the break, the stronger field they produce. It is a natural phenomenon in nature. The finer they become, the finer frequencies they produce which is exponentially more powerful than any frequencies available described in the string theory. Maharishi said that they are sounds of natural law reverberating in the whole universe as they similarly reverberate silently within the human physiology.

Maharishi referred to it as the "sap" of the universe, I called it the AS_U particle, the anima solo particle that nourishes and animates the whole universe. Maharishi said that it creates "from itself, by itself, within itself," so, it must be a solo particle, not like the other elementary particles of nature, they come in pairs, with a neutron, with a quark or with an anti-particle. The AS_U particle, is a solo

---

[28] *See Chapter 6, Global Taste of Utopia*

[29] *See Chapter 7, The Science Behind Climate Change*

particle that comes by itself and it is the positive particle of all creation because thru recent validated experiments, they showed and created only positive trends, no negativity exists within its invincible field, within the Unified Field. By nature, the Unified Field creates by itself the purest positive particle in all creation. Like the photon particles create only light, no darkness can exist within its photon field. Light is the nature of the photon fields. Scientists now call these finest frequencies as strings instead of particles or waves. They are unseen, unmanifest and unobservable by the naked eye. Invincibility comes from the field they produce, the Unified Field, Maharishi referred to as the seed form of all the laws of nature. It can be enlivened by the human mind at the level of silence thru the practice of Transcendental Meditation.

Experimental model can be established in the Philippines and because the country is constantly visited by typhoons and is a Third World poor country, it will display the observable transformation easily more than any country in the world. It will display less typhoons occurring, in contrast to more than two dozen typhoons a year occurring. When the super-radiance group is established on a permanent basis, typhoons will completely fade away if not significantly reduced. Economy will rise to heights never before reached by the country, like the 19% economic increase in Mozambique, Africa in 1994.

The government is the beneficiary of the tremendous benefits created by the group practice of the TM and the TM-Sidhi or Yogic flying program. The Bolinao school campus property that I had bought in 1994, it can be used there again. The government should lead the program and invite scientist to monitor and document the transformation.

The significance of the experimental model in the Philippines is to remove the heavy burden the government undertakes evacuating and housing large populations displaced by typhoons year after year. With the result of the events in 1979 and 1984, twice repeated, Maharishi endeavored the super-radiance effect of the Unified Field be established in the Philippines to radiate to the world, like the sun radiating its rays from east to the west. The number required for super-radiance to occur is the square root of one percent of the population of the Philippines is about 1,500 participants formed on a permanent basis will completely reduce or even end wild typhoons in the Philippines.

Maharishi said, "The Philippines, the land of the rising sun, it will radiate happiness and fulfillment to the world." Maharishi was not well-informed. Japan is the land of the rising sun. The Philippines is the Pearl of the Orient.

In California, constant droughts prevail. Nearly two dozen of massive wild fires occur every summer. The cost of infrastructures and homes lost is staggering. Just about 600 participants gathered in groups together can relieve drought and end the ferocious wild fires in California. Transcendental Meditation has gone mainstream in the United States, I wonder why a group is not being established in California. There is a group in Fairfield, Iowa, but it has not reach the number of the 2,500 participants required for the United States to affect the whole country. TM and Yogic flying appears unconventional as a tool to solve problems of global warming, but rather than losing many homes and spending large amounts to rebuild burned infrastructures, it needs to be considered. It is well worth it for the State or insurance companies to review existing cooling evidences affecting climates made by the Unified Field of natural law in order to save money on rebuilding burned structures and further avert fires and droughts in the future.

Experimental models for climate change should be made on a permanent basis, it is a must do in our generation. For typhoons, the Philippines is an ideal place, typhoons are prolific. When the group of 1,500 TM and Yogic flying is place in groups together in the Philippines, it is easy to observe the great reduction or elimination of typhoons.

California has the most fires and droughts in the USA, as soon as 600 TM and Yogic Flying group is formed at the central part of California, less massive wild fires or none will be observed. Droughts will completely be eliminated. There are constant floods in Texas and along states bounded by the Mississippi river like Missouri and Illinois, they are overcome with constant floods. The historic snow storm that recently occurred thru Washington D.C to New York in January of 2016, it brought thirty inches of snow that cripple entire cities and flooded New Jersey. Nearly thirty people died. Establishing Maharishi' Unified Field group can significantly reduce or remove the great cost of damages floods create.

For the removal of poverty, the United Nations should establish a experimental model on a permanent basis in the poorest country in the world like Malawi, Liberia, Congo or Madagascar. Their economies will rise like those shown in many experiments done around the world.

In conflict areas, like Syria, Yemen, Iraq, Afghanistan, Africa and other terror ridden places, the invincible character of the Unified

Field of natural law will be observed when the square root of one percent of each country's population practice TM and Yogic flying in groups together. It will counter terror and end conflicts.

Nearly five years of war in Syria has caused the rise of millions of refugees fleeing the country, just a small number is needed to end the war. In Syria, it just needs less than 500 TM and Yogic flying participants formed, compared to millions of the population fleeing the country. With all the millions of dollars spent daily in the air strikes and the danger of rising military weaponry from Russia in Syria, it is much easier to form the 500, instead of further complicating and extending the whole scenario of war and human suffering in the country. During these long years of war and suffering in Syria, nothing successful was done today to avert the tragedy of millions fleeing refugees who bear the brunt of war. The United Nations could not find a solution. The coalition of Allied groups with the West are doing regular air strikes, yet ISIS seems not abating soon. When nothing else works, introduce a third option, just like in any experiment, introduce another element until the result is obtained. Introduce Maharishi's natural law principle of invincibility on a permanent and the results of peace will be obtained.

In the year 2200, it was predicted that the amount of devastation that will be caused by global warming will be running to $336 [30] trillion dollars. It will far exceed devastations caused by war. When the CO_2 layer further increase in concentration above the earth, it will increase absorbing heat that will warm the earth's atmosphere that will further melt the ice caps in the North Pole causing communities to submerge, dry up food crops on land with droughts and will drastically break the cold and warm air equilibriums causing wild winds, typhoons and hurricanes of great magnitude to occur on earth.

Particle physics is a recent find. A coalition of prominent scientists who have deep understanding of science, they can further research and monitor the creation of experimental models for climate change. In their hands lay the future of humanity. The human physiology has a natural built in temperature sensor mechanism, like when warmed, the sweat glands are activated and drips of water are formed in the human skin. Furthering the natural character of water, it means that when air is cooled down, it turns liquid water, when further cooled liquid water turns to ice at zero degrees centigrade and when further heated at 100 degrees centigrade, water turns to air, yet

---

[30] *Nature Climate Change Journal, September 1015,*

the atomic composition of water remains the same H_2O in all its three states, liquid, matter and gas, like water, ice and air. The water droplets that emits from the human body when warmed is a strong evidence that the human physiology posses the natural cooling mechanism nature has designed for man to be an instrument to exponentially cool the environment in large groups together with the application of the Unified Field. Like more magnets joined together, greater magnetic field is produced, more cooling mechanism from the human physiology increased in large numbers together, it cools the atmosphere. The solution to global warming is held by the natural cooling character of man. This is the one reason why Maharishi said, "Man becomes master of natural law!" Man cools his own body as he can cool his own planet earth in large groups together. Man is part and parcel of the planet, the constitution of the human body is the same constitution of the planet earth and the whole universe at the level of the Unified Field!

John Hagelin, PhD is a brilliant particle physicist, a Unified Field scientist and a string theorists, Maharishi appointed him to lead the TM movement in the United States, (see tm.org). Tony Nader, MD, PhD was named Maharishi's successor and Maharishi gave him the title, Maharaja Adhiraj Raja Raam and First Ruler of the Global Country of World Peace, (see globalcountryofworldpeace.org), he is also a prominent medical doctor, a brilliant neurologist and brain scientist.

There are many brilliant scientists around the world, from countries like Germany, Great Britain, France, Japan, the United States and other nations. They should work together to create and monitor the results of the experimental models for climate change, then apply it to each country and to the world.

There is no instant and immediate solution to global warming currently occurring, whatever or otherwise without mitigating economies of nations. No religion, church or invoking the name of any god can stop global warming. Only science can. Finding a scientific solution has been going on for decades, no solution has been found that can provide instant and immediate cooling effect to solve climate change, like pouring cold water in a boiling pot, the hot water cools down in an instant. Introducing the natural cooling character held by the human mind in large groups together at the level of silence, recently brought to light by Maharishi Mahesh Yogi, a physicist and a revered Indian master, thru the technology of the Unified Field, it has proven scientific evidences already peer-reviewed that now exist, see www.globalcountry.org.

Soon as large number equal to the square root of one percent of the population of each country or the world practicing TM and Yogic flying in large groups together is formed, the result is immediate and instant. I witnessed the instant cooling effect, not once, not twice but many times during the times I established large groups of students together practicing TM and with the workers on the job building the thirty storey tower (see photo of tower on Chapter 14) practicing TM before work and after. The cooling effect of the Unified Field held by the human mind at the level of silence in large groups together is immediate and instant. The cooling effect in 1994, in Mozambique was also instant. In Brazil in early 1984, foretold by Maharishi at the Global Taste of Utopia event was also instant. I was not the only one who observed the natural instant cooling effect held by the human mind in large groups together at the level of silence.

It is a great challenge for any government to duplicate it, like that done by then President Joaquim Chisano of Mozambique in 1994. The only problem when the leader of government is replaced, the following new leader may not have that open-minded view to continue the program, just like what happened in Mozambique. A committee of scientists endorsed and funded on yearly regular budget, just like any government program, it will better succeed in securing continuity of the program. There are also mega donors in the current political parties in the US with millions of dollars available to support initiatives for climate change, they too can sponsor similar experimental models or support already existing groups like the project at the Brahmasthan of India or the community of TM and Yogic flying practitioners currently in Fairfield, Iowa, USA. I have established the Maharishi Memorial Foundation for the same purpose, to form a college again, a Maharishi College or something.

Maharishi has foretold that all negativities in the world will fade away with the application of the Unified Field. The formula for the creation of heaven on earth has been brought to light from the knowledge held by the ancient Vedic civilization of India thousands of years ago, now interpreted with the discovery of modern science of the Unified Field of natural law. The human mind in large groups together, at the level of silence is capable to solve global warming.

10

VEDA AND THE ANCIENT KNOWLEDGE

The Veda contains the verse that describes the group character and behavior of the finest elementary particles of nature held by the human mind at the level of silence which states, "Go together, speak together. Know your minds to be functioning together from a common source. In the same manner as the impulses of creative intelligence in the beginning remain to be together near the source. Integrated is the expression of knowledge. An assembly is significant in unity. United are their minds while full of desire. For those who make use of the integrated expression of knowledge, by virtue of unitedness and by means of that which remains to be united, perform action to generate wholeness of life. United be your purpose, harmonious be your feelings and collected be your minds in the same way as all the aspects of the Universe exist in togetherness, wholeness."................Rig Veda.

Veda means knowledge. It originated from the ancient Vedic civilization in India more than 6,000 years ago. Because of the significant knowledge referring to human transformation contained in the Vedic literatures, Maharishi gathered and compiled these ancient books and had them interpreted in the format of modern science. Maharishi integrated these Vedic norms into a complete science now available at the Maharishi University of Management in Fairfield, Iowa, leading to a doctorate degree in Maharishi's Vedic Science.

Throughout history, many theories were written illustrating dogmas which claimed their ability to improve man and his surroundings, but facts proved that none has been successful as those that originated from the ancient Vedic civilization of India, particularly the Transcendental Meditation contained in the ancient Vedic literature, the Rig Veda.

The most significant interpretation came from Tony Nader, MD, PhD, under Maharishi's guidance, he discovered that the Vedic literatures are expressions of the human physiology. For his discovery, Maharishi gifted Dr. Nader his weight in gold.

Maharishi inaugurated the year 1994, as the year of the discovery of the Veda and Vedic literatures as expressions of the human physiology. It coincided with Tony Nader's book launching of the "Veda and Vedic literatures; Expressions of the Human

Physiology[31] that came out of the press. Maharishi named Tony Nader, first Ruler of the Global Country of World Peace and supreme leader of the worldwide Transcendental Meditation movement. Dr. Nader says that the structure and sounds of the Veda corresponds to the expressions of the human body and contains the same expressions of natural law as it emerges from the unmanifest, non-material form characterized by the silent frequencies inherent in the human nervous system as it manifests to material forms, to cells, blood and different parts of the body. Modern medical machines prove the fine frequencies of the human body, like the EEG measures the frequency of the heart, the EKG, the frequency of the brain and other equipments like the ultra-sound machine or the simple stethoscope. The Unified Field describes in modern format the sounds and frequencies of the Vedic literatures that prevail unseen and unmanifest in the human physiology before cells and matter forms of the human body begins, just like Maharishi referring to the "sap" of the universe that creates "from itself, within itself and by itself." The forty aspects of intelligence in the human physiology correlate with the forty branches of the Vedic literatures. Man can now apply these norms in life and gain peace, prosperity and create heaven on earth.

The ancient Vedic literatures held and contained the significant norms which remained hidden, missed and remained forgotten for many centuries. It described the secrets that man has the birth right to happiness and fulfillment, in contrast to the many problems and suffering man has faced for many generations.

The discovery of the Unified Field brought to light by modern science found that the ancient Vedic literatures are expressions of the whole human physiology. Several scientists confirm the physiological and neurological correlates of the Veda and the Vedic literatures referring to the functioning of the human mind and body aligning with the same sounds of natural law reverberating in the whole universe. The ancient Vedic norms held the secrets to human health and happiness. Now the most advance knowledge of modern times, the Unified Field of natural law has been transformed in a scientific format verifiable by the human mind at the level of silence in large groups together.

Albert Einstein recognized the existence of the Unified Field when he discovered that all matter in the universe can be broken down to the finest subatomic particles. Einstein was the fore-runner of

---

[31] *Veda and Vedic Literatures; Expressions of the human Physiology, by Tony Nader, MD, PhD*

the hydrogen atom and stipulated that matter and energy spring from the same one unified source, the Unified Field. All matter and all manifest form of creation spring from a non-material, unseen, unmanifest root energy, the Unified Field. The Vedic literatures describe the unmanifest, non-material form of creation, it said it is the same absolute and inner structure of the reality of creation and functioning of natural law in the whole universe. It has been brought out into the world and referred to as the theory of everything or the superstring theory described by modern physics operating at the level of the absolute unmanifest level of creation also found in the human mind at the level of silence referred to as the Unified field of all the laws of nature. The Unified Field has become the most advanced scientific knowledge of our time.

The sounds of the Veda and the Vedic literatures were passed on and preserved by families from the ancient Vedic civilization of ancient India in order to keep the valuable wisdom that they contain will never be lost. In the ancient Vedic civilization that was once prosperous, it kept perfect the structure of the Vedic literatures as a tradition passed on from generation to generation. However thru the many centuries that passed in time, foreign cultures mixed with the population and in the attempts to conquer India, the Vedic literatures became fragmented and scattered. Some of its true meanings were lost due to different interpretations and different traditions that came about and obscured their fundamental truths. Soon the perfect sequential structure of the Veda and the Vedic literatures became forgotten. Its importance to practical life was ignored and the significance for use in the life of people was obscured and forgotten.

Once the tradition was lost, people started to suffer and many problems in the general population sprouted. Soon life on earth fell into the state of suffering we have seen rising day by day. Crimes, incurable diseases, floods, typhoons, earthquakes, fires, wars and economic downtrends are negativities that plaque the population. All these negativities need to be phased out of the environment. For the first time in the modern world, a formula has now been found as a solution to these problems.

Maharishi found the solution to mankind's maladies from the ancient knowledge of the Veda and the Vedic literatures where no modern technology has solved many problems of humanity that beset modern man. Maharishi related the wisdom from the ancient Vedic literatures to the modern science of the Unified Field, the unmanifest, non-material aspect of nature that the Ancient Vedic tradition extolled. The Vedic literatures are expressed in the human physiology and within the human mind, it can now effectively process the creation of heaven on earth. Maharishi has gathered and brought to light the priceless wisdom from the forgotten heritage from the ancient Vedic civilization. At the level of unity consciousness, Maharishi's Vedic Science came from Maharishi's cognition and interpretation of the Vedic literatures as the basis of the human physiology.[32] He once again brought back the wisdom from the ancient seers and masters of the ancient tradition of the Veda and Vedic literatures from the long lost antiquity of time. Not only did Maharishi cognized and interpreted the ancient knowledge of the Veda and Vedic literatures to significantly affect the well being of man and his environment in the creation of heaven on earth, his whole life and self referral experience was dedicated to the propagation of these ancient Vedic norms and in the search for its modern interpretation working hand in hand with modern scientists interpreting the Veda and Vedic literatures and related it to modern science that led to the discovery of the Unified Field of Natural Law.

The impulses of the Veda are one and the same impulses of natural law at the level of the Unified Field. The primordial sounds of the Veda are the faintest vibrations of Natural Law at its seed form perpetually residing at the Unified Field and conducting the evolution of the universe. It is the seat of the fundamental laws of nature organizing the activity of the universe in perfect harmony and order. The sounds of the Veda are sounds of natural law.

The Vedas contain the sound prevailing in the universe and creates the same harmony and order found in the human mind and physiology that is why the level of the Unified Field available in the human mind is at the most supreme and fundamental wisdom that affects the entire well-being of humanity.

The Unified Field is the seat of nature reverberating in the human mind that traverse and provide balance to the whole human physiology. The Vedic literatures were found to be the basis and are the expressions of the human physiology. This historic discovery has brought to light the supreme benefits that can be attained by humanity in this generation. Maharishi has brought to light the Veda and the Vedic literatures to bring about life without suffering. Even in his death bed Maharishi said, "I have completed my duty to Guru Dev." Maharishi has fulfilled the cause of man to its highest most supreme state of the evolution, a heaven on earth for all men to enjoy.

---

[32] *Appendix D, Science of Being and Art of Living, Revised and updated, November 2001*

Dr. Tony Nader, MD, PhD, a brilliant scientist with Maharishi's inspiration discovered and confirmed that the Veda and the Vedic literatures are fully and completely expressed in the human physiology. In his book "Human Physiology: Expression of the Veda and Vedic Literature,"[33] Dr. Nader described the forty aspects of intelligence at the basis of the human physiology.

Dr. Nader is no ordinary scientist. After graduating from a medical degree at the American University in Beirut, he pursued his PhD. in Brain and Cognitive Science from the prestigious Massachusetts Institute of Technology and became Assistant director at the Clinical Research Center.

He completed his post-doctoral work and became a Fellow, the highest professional recognition in the field of neurology at the Harvard Medical School. Furthering his quest for total knowledge on the brain and human physiology, he studied Maharishi's Vedic Science and Technology under the guidance of Maharishi Mahesh Yogi. Based on his advanced studies on the brain and human physiology, Dr. Nader has successfully correlated each aspect of the Vedic literatures to specific area of the human physiology with the conclusion that the human physiology is indeed the expression of the Veda and the Vedic literature which finally brought to light the wisdom from the ancient Vedic civilization that Maharishi wanted modern science to confirm.

Maharishi gave Dr. Tony Nader the supreme title of Maharaja Adhiraj Rajaraam, named after King Rama who once ruled during the prosperous ancient Vedic civilization of India. His recent book, Ramayan in Human Physiology describes the ancient book Ramayana not only is an epic from long ago but the Rama, Ramyana and the Ramraj is found literally in our own nature. It is the eternal cosmic story, the Unified Field of Natural Law unfolding in each and every human being. Maharishi described "Maharaja Ahiraj Rajaraam, (Dr. Nader) is talking in terms of the mechanics of transformation of Natural Law into physiology. He has realized that the total Constitution of Natural Law, which governs the universe in perfect order, is lively in every grain of physiology. The language of the Ramayan is that language in which total Natural Law is actually seen administering the whole universe."

Maharaja Adhiraj Rjaraam, Dr, Tony Nader described the supreme version of the Veda and the Unified Field lively in every human being. The characters of the Ramayana express the character

---

33 *Maharishi Vedic University Press, Vlodrop Holland, 1994*

and functioning of the human physiology that took form in scientific terms. His brilliant interpretation of the Vedic literature as the expression and vibration of the Veda, the total knowledge of natural law occurring within the human physiology defines man's connection to the universe and creation. Every aspect of the human physiology is the finest vibration of natural law and is the same sounds that create the whole universe in perfect harmony and order. That is why Maharishi's explanation of the ancient Vedic literatures and Dr. Nader's correlation with the human physiology truly confirm that man's true birthright is harmony and order similar to the orderly functioning of creation and the universe which can bring fulfillment in life in tuned with natural law. Natural law operates in perfect order and harmony, it should function in similar order and harmony inside the human mind and body in the same orderly functioning of the universe.

Dr. Nader's discovery of the Veda and Vedic literature in human physiology is truly significant in the implementation of Maharishi Ayurveda in the enhancement of the healing ability of the body described at the finer more subtle level of the human physiology. Dr. Nader's researches in neurochemistry, neuro-endocrinology and its relationship with diet, age, behavior, moods, seasonal influences, neurotransmitters and hormonal activity are precursors in medicine that perfectly supports Maharishi Ayurveda.

Dr. Nader's historic discovery is being put to use at the Raj in Fairfield. The knowledge is processed in precision to the detail in every aspect of treatment at the Raj in Fairfield, Iowa, that is why 50% relief from chronic diseases and well being is gained in just five days of treatment. For this reason, modern medicine can work hand in hand with Maharishi Ayurveda in the prevention and treatment of disease at the very root of the symptoms.

Modern medicine targets the material, seen and felt aspect of the human physiology where the symptoms of the disease occur. Maharishi Ayurveda, in line with the ancient Veda and Vedic literatures targets diseases at the basis of the unmanifest, non-material, finest vibration frequency where the source of the ailment begins, at the Unified Field occurring at the silent level of the mind, thus, consciousness is the main center for remedy in every human malady.

An ancient Rishi, a Vedanta, a Guru Dev, a Maharishi incarnate and a King Rama has risen in our time. Dr. Tony Nader deserves the greatest honor, one who both is expert in the ancient Veda in human physiology and expert in modern science of the

Unified Field. Now, Dr. Tony Nader, MD, PhD has become the first ruler of the Global Country of World Peace and guardian of the Vedic tradition in line with the tradition of the ancient Vedic masters. He has become Maharishi's embodiment of the Veda and Vedic literature.

The knowledge of the Veda expresses the qualities of the Unified Field, the AS_U (anima solo particle of the universe),[34] the primordial particle interacting with the Unified field, like the graviton particle interacting in the gravitational field, electron particle in the electric field or the magneto particle at the magnetic field.

The Unified Field refers to the knower, the process of knowing and the known value in human consciousness described in Rk Veda, as the Rishi, Devata and Chandas and their unity called Samhita or the Unified Field that creates "from itself, by itself, within itself" is the eternal structure pervading in the whole universe. This seed form of nature organizing the perfect harmony and order of the universe is similarly found within the human mind at the level of silence during the practice of Transcendental Meditation.

Dr. Nader illustrated the expression of the Veda and the Vedic literature and its forty qualities of intelligence at the basis of the human physiology. The Physiology Chart attached displays the significance of the fine pathways by which the particles at the level of the Unified Field, the AS_u, particles that traverses the whole physiology. The sounds and the fine frequencies produced by the Veda and Vedic literatures trace the fine pathways of the human physiology thereby balancing the rhythms of the body cell along its path. Since the human physiology is the conductor of the Unified Field, the sound of the specific Vedic literature strengthens, sharpens and targets the specific area of the physiology. Like the *Sama Veda*, it sharpens the functioning of the sense organs, like hearing, sight, touch, smell and taste. The *Upanishad,* the physiology of the ascending tracts of the nervous system and the *Brahmana,* the descending tract of the central nervous system. The S*mriti* sharpens memory.

Just like in a car, the electron particles trace the pathways designed for the car to smoothly run like the motor, the carburetor, the radiator, the transmission, or the exhaust system among them. All the parts participate in the particle pathways and enhancing the specific parts for ideal functioning and balance of the system, then the car moves forward at specified speed.

---

[34] *Chapter 7, The Science Behind Climate Change*

VEDA AND THE VEDIC LITERATURES AT THE BASIS OF THE HUMAN PHYSIOLOGY

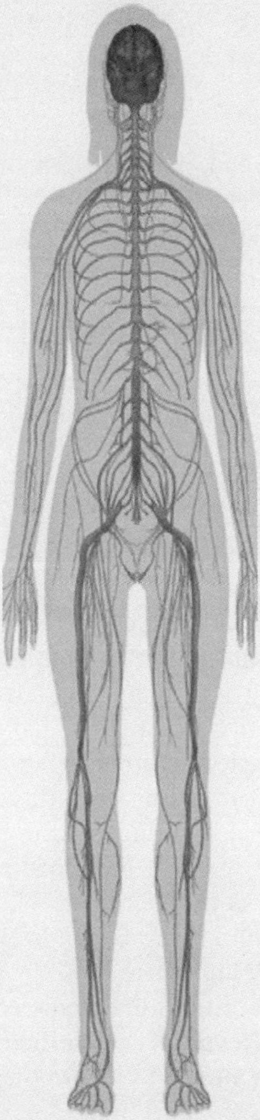

THE PHYSIOLOGY CHART

It is similar to the air path taken in the air-conditioning system. The air has specific pathways, the air intake, the motor, the exhausts, like that, to function perfectly in the balancing the air. It traverses every single part of the system so that the air becomes warms if the room is too cold or it cools the room if it is too hot with the correct control.

Similarly, the ASu particle traverses the whole human physiology and every center of the brain, and referring to global warming, brings about the character of cooling the atmosphere at the level of silence in large groups together. Referring to peace, the ASu particles traverses the whole human physiology and every chamber of the brain that brings about the character of kindness and compassion that eliminates anger that causes conflicts and war. Referring to eliminating poverty, the ASu particles traverse very cell of the human brain that brings about one hundred percent of human intelligence, and unfolds knowledge in accord with natural law that brings about infinite creativity needed for economic prosperity. The ASu particle nourishes every system of man and his environment, like a tree, it waters the root of all life and creation. When gathered in large groups together, it balances climates, creates peace and prosperity on earth.

The human physiology in itself is the most complex system of creation. It has the nervous system, the respiratory system, the digestive system, the lymphatic and endocrine system and the circulatory system to name some basic system of the body. The human body needs ideal balancing to be truly functioning in accord with natural law. The forty aspects of the Veda and Vedic literature[35] describes and illustrates the hundreds of pathways the $AS_{U,}$ has to traverse and bring balance to the functioning of the whole body. The finest anima solo particle of the universe, traverses the forty branches of the human physiology, also enumerated in Dr. Nader's book, "The Human Physiology, Expression of the Veda and Vedic Literatures." Man then lives a life that is problem free. This is the significance of the Veda and Vedic literature in human physiology.

The ancient Vedic literatures hold the ancient secret of health and happiness of man. It balances the whole physiology and awakens the finest particles at the level of the Unified Field occurring at the silent level of the human mind. For these reasons Maharishi emphasized the significance and importance of the reading and recitation of Vedic literatures, parts of it daily.

---

[35] *The Human Physiology Expression of the Veda And Vedic Literatures, by Dr. Tony Nader, MD, PhD*

The Transcendental Meditation is part of the Rig Veda that transforms the whole physiology that is scientifically found to give rest to the body twice deeper than sleep in just twenty minutes and provides balance to the whole physiology and reduces stress.

Herein enumerated are significant neuro-physiological correlates at the state of transcendental consciousness, a state of higher consciousness experienced during the practice of Maharishi's Transcendental Meditation. The following correlates show the changes occurring in the body in just twenty minutes of TM practice.

The forty basic physiological functioning of the body is strengthened, enlivened and balanced by the text, verses and sounds of the different ancient Vedic literatures. The sounds and frequencies of the Veda and Vedic literatures expresses the totality of the Unified Field which traverses the whole human physiology producing a balancing effect needed for the well being of the individual to fully act as the conductor of the Unified Field, that way, the individual lives in accord with natural law and gets support of nature throughout his life.

In 1977, Keith Wallace, MD provided a comprehensive research on the effect of the practice of TM in the mind and body and its neuro-physiological correlates and published his researches in several volumes of the *"Scientific Researches on Transcendental Meditation."*

Dr. Keith Wallace researches have been peer- reviewed and repeatedly validated by more than 600 prestigious universities in the world that is why Transcendental Meditation is now being recommended by many doctors for the well being of individuals and prevention of diseases.

NEURO-PHYSIOLOGICAL CORRELATES OF TRANSCENDENTAL MEDITATION[36]

Electro Encephalogram
(Brain Wave Functioning)

- Awake – low voltage mixed beta and theta wave frequencies and/or alpha wave activity. Alpha wave activity increases with the eyes closed. A moment of tension or an attempt to solve a problem may disrupt it.

---

[36] *Excerpts from Scientific Researches on Transcendental Meditation, Dr. Keith Wallace, 1977*

Tiredness or drowsiness is characterized by a flattening of the alpha waves to a low voltage mixed wave frequency with an increase in activity.

- Sleep – decrease in regular wave activity.
- Dreaming – low amplitude variable frequency waves.
- Transcendental Consciousness – increases the alpha wave activity that occurs in the central and frontal regions of the brain indicating alertness. More refined theta and delta waves.

Electro-oculogram

(Eye Movements)

- Awake – rapid eye movements, eye blinks, various movements
- Sleep – first slow eye movements then eyes role up and become immobile.
- Dreaming – episodic rapid eye movements
- Transcendental Consciousness – no apparent change.

Electro Myogram

(Muscular functioning)

- Awake – changes according to activity.
- Sleep – less than during wakefulness
- Dreaming – head and chin muscle relax. Tonic mental and sub-mental activity reaches low level.
- Transcendental Consciousness – no change in muscular activity recordable.

Body Movement

- Awake – changes according to activity.
- Sleep – cannot stay erect; motion uncommon (except in somnambulism) must lie down or sit to support spine in sleep.
- Dreaming – relaxed although spasmodic movement of certain muscle groups does not occur.
- Transcendental Consciousness sitting comfortably, body becomes relaxed.

Oxygen Consumption (Intake of 0_2)

- According to activity, oxygen intake changes.
- Sleep – in 6 to 7 hours of sleep, oxygen is reduced to 10 to 20%.

- Dreaming – consumption of oxygen is higher rate than sleep state
- Transcendental Consciousness – decreases significantly by mean decrease of (16%) in 15 minutes. It is greater than most values reported for a mean decrease over a full night sleep.

Body temperature

- Awake - highest value of circadian cycle is attained. Circadian are brain cycles as the sun orbits the earth
- Sleep – lowest value of circadian cycle is attained
- Dreaming – rhythms varies constituting sub-cycles or modulations of circadian cycle
- Transcendental Consciousness – sub-cycles considerably reduced.

Cardiac Output

(Heart Beat)

- Awake – faster or slower according to activity.
- Sleep – about decrease 20% in 6 to 7 hours of sleep. During deep sleep, the heart rate, respiration rate, blood pressure fall to the lowest level of the circadian cycle. It falls at sleep onset and continuing to drop until one hour before waking.
- Dreaming – Usually 5% decrease over deep sleep, erratic.
- Transcendental Consciousness in 15 to 20 minutes heart beat decreases significantly with a mean decrease of about 25%. Some significant decrease of about 5 beats per minute.

Blood Gases

- Awake – changes according to activity.
- Sleep – increase of PCO_2 significantly and significant decrease of pH.
- Transcendental Consciousness –not significant decrease in pH. Base excess significantly decreases.

Blood Pressure

- Awake – changes according to activity.
- Sleep – reduces during deep sleep.
- Dreaming – erratic. Usually decreases about 5% over deep sleep.
- Transcendental Consciousness – blood pressure readings decreases of about 3 beats per minute with a significant mean decrease.

  Change from 12 to 4 beats per minute.

Skin Resistance

- Awake – varies according to activity.
- Sleep – there is an increase of about 50% in skin resistance.
- Dreaming – unchanged, not significant.
- Transcendental Consciousness increases significantly. In some cases there is a 500% increase of after only ten minutes of meditation.

Blood Lactate Concentration

- Awake – with stress or activity it increases especially in patients with anxiety neurosis.
- Sleep – decreases
- Transcendental Consciousness – significantly decreases.

Awareness

- Awake – changes according to activity.
- Sleep – it is hard to awake a subject with a low noise or buzzer that would have aroused him earlier. Subject usually awakens slowly and may feel not experiencing any mental activity.
- Dreaming – a relatively loud noise is required to awaken the subject, although a very slight noise with significance may quickly alert them.

- Transcendental Consciousness – attention is primarily attracted inward process of thought. Responsive to significant or disrupting external stimuli. Expanded inner awareness both during and after the practice of Transcendental Meditation. They describe the transcendental state as "restful alertness."

Sleep Deprivation

- Sleep – prolonged deprivation can produce sensory disorders, disorientation and loss of efficiency in physical activity. There is an alteration in EEG activity, energy metabolism, and the hormonal and biochemical conditions of the body after one to five days of deprivation.
- Dreaming – subjects become disoriented, irritable, hallucinogenic, and sometimes, psychotic. In subsequent nights, subjects increase their dreaming.
- Transcendental Consciousness – deep relaxation and rest are natural and necessary to the proper functioning of the body

and mind. Transcendental Meditation gives a deep state of rest with minimal effort.

- Deprivation of deep rest and rejuvenation is postulated to be the cause of physical diseases and mental diseases.
- Deep file up of stress and physical fatigue can be released thru the practice of Transcendental Meditation and prevent disease.

Thru Maharishi's guidance and inspiration, many modern scientists has brought the Veda and the Vedic literature into a scientific format and they interpreted the Unified Field of modern science in a more understandable layman's language. Maharishi was a physicist and a monk, he is considered the greatest scientist of them all in the field of mind sciences and consciousness.

Sometimes in 1986, John Hagelin, PhD, a summa cum laude from Dartmouth College and a PhD from Harvard University together with Maharishi illustrated and described the Unified Field underlying the deepest and finest level of creation that Einstein pursued and failed to locate. The discovery of the Unified Field brought to light the most supreme of all sciences referring to the total transformation of man and his surroundings. In recognition for Dr. Hagelin's outstanding achievements, he was named the winner of the Kilby Award in the tradition of scientists like Einstein, Jeans, Bohr and Eddington. Maharishi named him the National Director of the TM Movement in the United States which he holds until now.

Maharishi confirmed the title of Maharaja Adhiraj Rajaram, to Tony Nader MD, PhD. Under the guidance of Maharishi, he has brought to light the expression of the Vedic literatures in the forty qualities of intelligence at the basis of the human physiology.

Dr. Tony Nader's two books *"Human Physiology: Expression of the Veda and Vedic literature."* and the *"Ramayan in the Human Physiology"* unfolds the mechanics of the functioning of the human body at its deepest and finest level that is truly the basis of perfect health. Science has now the means to interpret the secret wisdom the ancient Vedic literature foretold by ancient Rishis many thousands of years ago.

Maharishi discovered and also expressed the benefit of the TM-Sidhi or Yogic flying in Patangili's verse;
Tat sanidhau vairatyagah....(Patangili's Yoga Sutra 2.35)

"In the vicinity of the phenomenon of Yoga, Yogic flying, conflicting tendencies disappear. The unifying influence neutralizes the diversifying qualities in the area."

Automation of government administration is expressed in the ancient text as follows;

Yatinam Brahma bhavati sarathih…..(RK Veda 1.158.6)

"For those who are established in the singularity of fully awake, self-referral consciousness, the total potential of pure knowledge and its infinite organizing power - the lively Constitution of the Universe, Natural Law-becomes the charioteer of all activity."

On Guru Purnima Day in October 2000, Maharishi established the Global Country of World Peace, a country without borders for all peace-loving citizens of the world. Maharishi described Administration through Natural law. He said; "It is a very, very great delight for me that I have a chance now, after Puja[37] to Guru Dev, to express to you all, for the world of all times the establishment of this global organization which has now been established with the blessings of Guru Dev as the Prasad of Guru Puja. "This global organization will establish administration through Vedic Knowledge in the center of every country, and this will enliven the organizing power of Natural Law for the government and give invincibility to the government, peace and happiness to the people and will give nourishing sovereignty to every sovereign nation."

"Now, through this global organization, I am inviting all governments who have been living under the influence of other nations to enjoy real freedom - no country in the world should be dependent on any other country: Get into this Knowledge of Natural law, enliven Natural Law in your people, enliven Natural Law in the collective consciousness of your nation and you will be the custodian of Natural law. This I am saying to every government: You will hold the balance of power in the world even if you are a small country. As long as you are a free country, as long as you have the power to make laws, as long as you have a currency, establish this center of Total Knowledge - Brahma Sthan - in the center of your country, and as knowledge has organizing power, the bliss of Total Knowledge – Brahmananda - lively in your country, the infinite organizing power of Natural Law will be at your command - and you will command authority in the family of nations - you will hold the balance of power in the world. My global organization that is being set today is to give sovereignty, real invincibility - unrestricted authority of sovereignty to every government."

---

[37] *Puja, the ancient Vedic ceremony of thanksgiving to the tradition of masters*

Maharishi Vedic City is the capital of the Global Country of World Peace and display the prototype of the Global Administration through Natural law in Fairfield, Iowa, USA. What is needed at the Maharishi Vedic City is to increase more participants in the Yogic flying. Housing developments are available in Fairfield and in place for people who wish to relocate there. There are also Vedic pundits from India that need financial support for their board and lodging to continue their stay in Fairfield. It is up to the people with means to provide invincibility for the USA soon as 2,500 is established in Fairfield.

Maharishi envisioned a super radiance group of 9,000 for the whole world at the central part of India. Maharishi, before he passed away, he instructed to organize a group intended for the 9,000 where nearly 1,200 have been established at the Brahmastan of India, see www.maharishivedicpundits.org. To gradually increase their population, there is a great demand for more sponsors and donations to feed them daily and house all the rest.

It is a huge responsibility to make the world a better shape now and in the next generations to come. Again, it is up to the people with means to make it happen. The cause for the good of humanity is the greatest and most supreme cause of Natural Law, it follows therefore that the reward for those who will support this great endeavor for the cause of mankind is exponentially the greatest of all.

Like a plant, you are watering the root to nourish the plant for it to blossom. The world will be ruled by the orderly and organizing power of Natural Law that is problem free. All men will enjoy heaven on earth.

11

MY SECOND TM SCHOOL

In 1988, my second TM school came about by accident. It all started when a TM practitioner dropped by my office one day. He said, "I have a piece of property in a prime area that I think you should buy and put up a school of Transcendental Meditation."

"I am not ready to buy any property now. I do not have the money," I said. I was expecting him to leave after my statement, but he stayed at my place moving quietly around, watching me go on with my tasks. He was persistent. He said, "You can give me a small token amount as a down payment and get the rest from the bank."

"Oh, the bank!" I exclaimed. I looked up at him from my desk, wondering about the possibility of getting the money from a bank that he was referring to. I remembered that my company had done work on some bank tenant renovations and I had become friends with one of the President.

Suddenly, I stood up from my table and told him, "Tomorrow, I will call the bank and talk to the President whom I know and let's see what happens."

I went to the bank the next day and explained to the President about my plan to put up a school. My company had done the construction management for the improvements of the bank's main branch, a twelve storey building located at the financial district of the city and just a few minutes from the property being sold to me.

"A school?" The President said wondering what wild endeavor I wanted to do. "Yes, a school, Sir. I put up an engineering training school before. I was just renting the place. Now, there is a place being sold to me, very cheap and I can put up a real college" I told him.

"If it is for a school, you will get the money," the President of the bank said. "Submit all the requirements, and you will get your school." The President called the loan manager and instructed him to take care of all the documents I needed to get the money for the school. It was such a surprise. There were no questions asked, just like that. It was so easy, in just one sitting, the loan was processed. In a few days, I got the money to buy the property for the school.

I remembered what Maharishi told me the first time I met him, "If your intention is pure, nature will support you." When you start to think in accord with natural law, your desire becomes evolutionary in the direction of all good. This is the fulfillment of desire described thru the practice of TM. Transcendental Meditation which allows the conscious mind to be in accord with natural law.

I started to experience the support of nature that Maharishi was talking about. When you serve and work for natural law and its cause, nature opens the door for you. You become master of natural law, nature fulfills your desire. It is not that you serve nature. Nature serves you. It is the reverse. "You become master of nature," Maharishi once remarked.

You command nature and nature fulfills your desire. This is the secret and greatest wonder of Maharishi's teaching. Serving the cause for the betterment of man is nature's cause, serving nature's cause, you become nature's master, very strange. I thought of the tale of Aladdin and the magic lamp. It is not Aladdin who creates the magic. It is the genie, the servant who creates the magic. He serves Aladdin, the genie's master. It is the servant that creates the magic, not the master. It is the reverse analogy of what we've commonly been taught to believe. It is not the master that creates, it is the servant. Reverse analogy at work! I would rather be the servant rather than the master, because then I can create the magic. Nature supports and serves the cause of man. It is like nature asking, "What is it that you wish Master? Your wish is my command, Master. I will do the magic for you."

Nature serves, not man, as long as man's cause is service to improve and support the cause of humanity, nature serves man. Maharishi illustrated his version, "You become master of natural law." Nature's cause is the cause for the good of humanity and if you follow nature's cause, doors will open wide to fulfill your desire.

This is one secret of Maharishi's teaching so few understand. "Desire," nature says, "and I will do work for you." Man becomes a channel of natural law at the level of the Unified Field. I always do my regular meditation and let nature run its course. I never miss it for any reason.

I got the school property and started to renovate the two-storey old dilapidated warehouse building after the bank released the money to pay for the property. It was an old wooden structure in a prime location in the city. It has not been used for more than a decade. We opened the main door and entered the warehouse. It was very dim with all the light switch not working and full of cobwebs like a scene in one of those out of a Hollywood horror films. There were piles of dust on the floor and several debris were scattered all over the place. We hurriedly gathered and disposed the scattered fixtures and debris. We fixed the light switches and replaced the old light bulbs.

Luckily there was a nearby city fire station. I went straight to the fire chief's office and requested if he can send some fire trucks to help us clean and wash the building. Again, with no question asked, the fire chief sent two fire trucks. They washed away the dust from the old warehouse.

There were spectators in the street curiously wondering, there was no fire burning, but why are there two fire trucks dozing water inside the old building. If only they knew, nature was supporting a TM school ready to open.

We had only two weeks to prepare for the opening of the school building. I hurriedly prepared a two-year associate vocational curriculum in engineering and construction. It wasn't very difficult because my engineering company had all the plans, and we were familiar with all the work described in the course outlines. The courses were designed for a two-year associate degree leading to Bachelors degree in Engineering.

I brought the curriculum for approval to the Department of Education in the downtown Manila office. I brought a team with me. My secretary and one teacher of TM named Rey. On the way out of my office, Rey was bragging to our staff, "The Secretary of Education is our boss's friend" pointing at me. His head high and waving his hand good- bye proudly to our staff as we climbed into the car. We came in to the Secretary of Education's office and entered the room.

The Secretary stood up from his chair as soon as he saw me. Appearing frightened, he said, "Don't you dare come to me for approval for your TM school. You are not my friend if you insist on teaching TM."

He hurriedly moved out of the room leaving all of us surprised and speechless. Rey was the first to come out of his unexpected stupor. He was a TM teacher who volunteered to work for the school.

Rey said, "I thought he was your friend. He was so scared of you like he saw a ghost!" I had to explain to my staff why the Secretary ran away. "He was a friend and a colleague in the structural engineering profession. We were both officers of the association at one time."

"He got so frightened because he was afraid to lose his job, poor guy. I do not look scary, do I? He was not afraid of me, he was afraid to lose his job. My colleagues in the structural engineering profession ridiculed me for becoming a Transcendental Meditation teacher and my intent to put up a TM school. What is a structural engineer like you doing with this kind of thing especially after the Catholic Cardinal Sin vehemently opposed TM in public and banned TM in schools. You know the breaking news with Cardinal Sin and all the controversy," I recounted.

"He was afraid of the Cardinal. His position is appointive at every regime term and one pressure from the Cardinal, his position will be gone." I continued to clarify the issue of the fleeing Secretary of Education to my staff who were with me. They were extremely frightened and disappointed by the incident.

Then, I suggested, "Let us go to the office of the vocational-tech department director. His position is permanent. I do not know him, but he might listen to us." The team hesitantly followed me to the next floor and to the office of the voc-tech director. We sat and waited outside of the director's room until we were called in. I glanced curiously at my staff looking troubled seated beside me. They had that blank stare as if expecting another similar unexpected episode was about to transpire and drive us permanently out. I placed my hand over my secretary's shoulder and raised my thumb up to show my staff some sign of encouragement as we were called inside.

I presented the curriculum I had made to the director. He read them page-by-page and occasionally stared curiously back at me, reviewing the summary of the courses. On occasion, he stared at me seemingly doubtful. To clear his doubts, I placed my business card on his table and said, "Sir, I am the President of the engineering company, our engineers will teach the students."

He took my card, seemingly unbelieving and scrutinized it slowly, then smiled. I felt so relieved. He said, "This is really good. Only two years and the students will learn actual construction techniques?"

"Yes sir," I replied confidently. "This is a good thing. It will help a lot of students make money earlier than the regular five year engineering courses. We will have plumbers, electricians, air conditioning technicians, estimators, construction managers and all the other skills we have prepared here in the curriculum, in just two years. We will reduce the brain drain of engineers going out of the country," I explained, convincing the Director.

"It costs so much to educate students and when they graduate, they leave for other countries to work seeking better income. Millions of our professionals leave, they become overseas-workers. This curriculum you have prepared will be of great value," the Director remarked. His remarks made me truly happy.

"This seems to be an initial step, an easy concept to start small businesses and increase job opportunities in our country," he was proudly smiling. I was flattered and I smiled back. At that time, way back in 1986, vocational-technical schools were not yet common in the country.

"We already have the expertise in place, as our engineers will be the one to teach and we will use the engineering plans that we used in the projects that we have already built," I included.

Then hesitantly, I bravely stated, "I plan to include Transcendental Meditation too." I anxiously waited for his comments, uncaring about the ongoing breaking news about Cardinal Sin and his objection to the teaching of Transcendental Meditation.

"What?" He stood up. My staff also stood up ready to move away and leave. Then the director said, "Aren't you afraid of the opposition from the Cardinal and the church?" Apparently, he was updated with the current breaking news and was aware of the prime time controversy, at that time over Cardinal Sin's front page objections with Transcendental Meditation. He became obviously worried.

"If we really want to fulfill the real purpose of education, in expanding the minds of the students, we have to do what we need to do to provide a well rounded overall educational experience. Besides, I have these three large volumes of scientific research proving that Transcendental Meditation is really good."

I brought out the three sets of blue-colored large volumes of books in anticipation for any objection. I was prepared to show them to the Director. They were the volumes of the "Scientific Researches on Transcendental Meditation, written by one of the scientist who worked previously with Maharishi. "These were written by an American medical doctor, Dr. Keith Wallace. Some research were funded by the prestigious National Institute of Health of the United States," I further elaborated.

TM was not popular in the country at that time in the early 1980. It became popular later after the Cardinal's strong opposition to TM and the population started to be curious about the idea of the Transcendental Meditation technique. Many press people learned TM and then they wrote good things about it. A couple of press people sought me out after Maharishi asked us to go on public TV at one time. I said in the TV interview when I was asked, "What is a structural engineer like you doing with TM?" Smiling, I answered,
"As a structural engineer, I design structures and buildings just so they do not fail. I became a teacher of Transcendental Meditation just

so people learn it and do not fail to be happy and healthy in their life. TM has a lot of scientific research what shows increase in intelligence, improved well-being and it brings happiness too." Then I raised the large blue volumes on the Scientific Researches on Transcendental Meditation for the audience to see. There was a loud applause from the audience. They were the same books I presented to the Director of Voc-Tech Education. The opposing Cardinal must have watched the TV interview and found out there was nothing wrong or controversial on the issue of Transcendental Meditation. I even sent copies of some research to the Cardinal thru his secretary.

The Director of Education took the books scanning the pages carefully and loudly reading paragraphs in some pages. "You really are ready to face the Cardinal," nodding his head. He confirmed, "The Cardinal is definitely opposed to Transcendental Meditation. It is well-publicized daily in the newspapers. Aren't you concerned about the Cardinal's strong opposition? And you want to open a TM school at this time and disregarding the Cardinal's opposition to TM?"

"Yes, I am prepared, sir," I said. "Besides, these scientific data is proof that Transcendental Meditation is good. Even Galileo was subjected to the inquisition of the Church when he postulated that the earth was round not flat."

"The church excommunicated Galileo for apparently tampering with the laws of God. We just need to demonstrate to the Cardinal that Maharishi's Transcendental Meditation has nothing to do with prayer or religion. It is a scientific endeavor that effectively transforms man and his environment."

"Galileo insisted until his death bed that the earth is round and that Ptolemy was in error for stating that the earth is flat. Scientists later eventually proved Galileo correct. I would like to state too until I die that TM is truly good," I smiled back. He was listening intently with my rebuttal. "Many scientists have proven TM to be already good. Several well known celebrities and millionaires in the USA have taken into TM. We have many current validations from prestigious universities. It makes students really, really intelligent," I emphasized. "Do not make the students too intelligent, otherwise the church will excommunicate you too," he said laughing.

"If people become too intelligent, the leaders will lose control of the people. If your Transcendental Meditation is good, I am with you," he stated, amused by my presentation.

"Go ahead. If you encounter some future problems later, call me, I am with you" he repeated proudly. He put a stamp of approval on my papers. "There you go," handing me the papers. I was surprised he stamped my documents "Approved" in one meeting! My happiness was overflowing! I resisted jumping with joy in front of him. Instead, I stood and bowed before him in humble gratitude. "Thank you so very much, Sir." I bowed again.

"And good luck with your endeavor," he said smiling, shaking his head. He stood up and shook my hand. He had that curious wondering look, curious that someone has a brave heart and who want to do an endeavor under obvious and public opposition from the church. He had that doubtful wondering look shaking his head in disbelief worried that a young lady like me, with no political clout, would attempt to teach Transcendental Meditation against the public mandate of Cardinal Sin, the mighty leader of eighty million Catholics at that time.

I expected my friend, the Secretary of Education would be my patron, but alas, he disowned me! The Director in front of me appeared extremely worried, at the very height of the controversy, someone like me dared to disobey the popular Catholic Cardinal.

"I do not like Cardinal Sin either," he concluded seemingly worried that I might be burned at the stake like Joan of Arc. He tried to comfort me, aware of the well publicized current events about the Cardinal's statements vehemently opposing TM.

Then he said, "Now, you have a friend here." He was pointing his finger proudly to his chest and again smiling. I left the director's office with my heart jumping with joy. I have an ally in the Department of Education! Again, I got support of nature in one sitting! Natural law provided me an ally who just became a patron and a friend today!.

"TM is a tool that supports nature's supreme purpose," Maharishi had said. Transcendental Meditation can easily support the transformation of man and his environment. TM makes man live in accord with natural law.

I bow down at the feet of my master and teacher, Maharishi. All the staff who came with me were overjoyed too. They were laughing and smiling from ear to ear, as we left the director's office.

We then placed an advertisement in the news papers, printed flyers and distributed them in churches on Sundays. We had them posted in many public places. Within two weeks, we had four hundred students enrolled. Nature did it so quickly!

It was a new course concept with only two years and with no entrance examinations required. This was a major change from the traditional courses which required some form of entrance examination. Most of the students who enrolled with us could not go to college as a two-year degree courses were not available in any colleges at that time. Students sought for us. It was a good thing for our school.

"No exams required," it said in the newspaper ads and flyers that we distributed. The two year vocational-technical college that I opened captured naturally all the students that had failed the required exams to enter any college. It was a new system of hands-on concept of a two-year associate degree courses that lead eventually to a bachelor's degree in engineering.

We partitioned the classrooms hurriedly in the evenings while the students filled them during the day. We also gave each student an IQ test before and after TM for reference. I included herein the amazing IQ test results that we conducted to trace the increase of the student's IQ in the following months. They are attached herein in the following pages.

A month after school opening, coincidentally it was my birthday, all the four hundred students started to practice Transcendental Meditation simultaneously together.

The predicted occurrence happened. The minute all the students closed their eyes in meditation, rain fell on the roof. Surprised, I looked outside and saw there were no dark clouds that foretold the rain. The rain made loud noises falling on the steel sheets of the roof. Strange, the rain stopped suddenly as soon as the students opened their eyes after meditation.

"You are getting a twenty-minute rain shower from heaven on your birthday," one of our staff said. "This is a physics phenomenon," I said. "It is a hot day. The meditation activated the cooling mechanism built in the human mind and cooled the warm air and produced the rain, balancing the climate, as Maharishi has foretold TM would do if done in large groups together."

"Meditation balanced the climate," I said in awe after the rain phenomenon that had just happened." Maharishi had promised the people of Brazil that rain would come during the Global Taste of Utopia experiment in 1984 as soon as the 7,000 was formed. The result is the same as we experienced today, it rained too," I said.

In Africa, it rained too in the drought areas in Mozambique in 1994, when large groups of the military started to meditate. We were just duplicating the natural phenomenon of balancing the climate with the rain falling in our school.

As it was my birthday, there was reason to celebrate. Everyone was happy. You could feel the happiness exponentially radiating from everyone. All my staff worked day and night completing all the school work that was needed to immediately open the school. They had accomplished so much in so short a time.

They were all joking and giggling with the students. After all the work we did tirelessly, we were compensated with this feeling of intense bliss reverberating inside all of us. We were happy all the efforts we have done paid off. In the beginning it was chaotic with many students rallying against TM. We taught TM one room at a time like a bar of magnet joined one bar at a time until all the twenty rooms filled with students all together meditated as a group, then the student rallies slowly subsided. In between rallies we had to call a police force to keep TM going on without interruptions. The police captain was truly sympathetic to my call for support because even before the school started, being a single mother myself, I related to the widows of policemen who were killed and gunned downed in the streets. I provided tuition fees for their children to go to school. It all went so well, in contrast to my first TM school, where all the students were taught TM but no one wanted to practice TM. It would not have been success without the help of the nearby police force just standing by along the school corridors. It was such an amazing event, even the policemen moving around the school were happily joking with the students!

"The mechanics to happiness," I thought. It was so real, this blissful, happy feeling! This is what Maharishi had said about the exponential effect of the group practice of TM. The four hundred students radiated so much happiness together. It was an awesome feeling of bliss inside that stays with you twenty hours every day.

The bliss particle, the seed of natural law lingering within every one became enlivened and was bubbling with bliss. The Unified Field, this phenomenon of nature, held the character of bliss. Experiencing the exponential multiplier effect of the four hundred TM participants overwhelmed the very core of our being with so intense bliss! The enlivened bliss particle stayed within us every day as the students meditated twice daily.

After the students began to meditate, I was invited to see Maharishi. There was an assembly of TM participants in the Netherlands and Maharishi was there. It was announced that a school was established in the Philippines and that there are four hundred students meditating. I was seated at the back end of the hall. Everyone applauded.

No one knew in the audience it was my school. My name was not mentioned. I joined everyone clapping happily behind the audience, joyfully clapping as if it was not my school.

Afterwards, one of Maharishi's top aides, Benny Fieldman, came over from the podium where he was sitting with Maharishi and walked down the long hallway back to the end of the hall where I was and said, "You did very well. Congratulations."

I felt humbled with just one person honoring me among the huge crowd. In my mind I did what I had to do and never did I expect any honor from the public or anyone. I was just so excited about the benefits it gave to the environment around us. The bliss effect reverberating inside was more than any honor one can receive. You just want to do the TM group again and again!

I was called to see Maharishi in private afterwards, "You did it all by yourself," Maharishi said smiling and handing me a bunch of red roses. Maharishi was beaming with joy!

"Maharishi, it was such a great pleasure. For the joy of it all, thank you for the opportunity to serve your cause." I answered. I saw Maharishi's eyes gleaming. He knew that same feeling I felt serving his cause. It was the same feeling of bliss he felt serving his master Guru Dev. It was nature's gift of real joy indeed!

Two months after the school opened, the students were given another IQ test. There was a ten percent increase in the IQ scores after two months. Five months later, there was more than twenty-five percent increase in IQ achieved among the three hundred-ninety eight students. Not all of the four hundred students had increased IQ. There were two of them who were not doing the TM program.

I called the two students privately and they said. "Our religion does not allow us to do TM. We want to go to college and since your school did not require any college exam, we enrolled here." I said," TM will not leave any child behind in increasing the intelligence, so you two should catch up with the rest, if you like to move on or be left behind."

One staff member said, "With this progression of IQ data, the students will have an IQ status of a genius within two years time."

I said, "We are just validating and repeating many other previous test reports. That is what all the other scientific experiments indicated and have published, noticeable increase in intelligence, we just saw how true they said it to be."

I noticed too, that gradually the student's behavior had improved. In the beginning, before meditation started, there were common fist fights and rivalries among the students. With just a few weeks of TM, there were no more fights recorded, very strange! The change in the school atmosphere was dramatic, from a noisy group to a more subdued learning environment.

One long time practitioner of TM came and brought his daughter to study in the school. "I want her in your school so she can practice Transcendental Meditation here. No other school here teaches it." I felt humbled by his choice because the facilities were rather austere being a rehabilitated old warehouse building.

The daughter came by my desk one day and told me, "My dad was right, I do not feel lonely anymore. I am feeling so happy and energetic, in contrast to how depressed I felt before I came here."

"It is awesome how this change can happen so quickly. It is the best technique to make you truly happy. And my grades too, they've gone up, surprisingly, not like before, my grades were really low." I took note of how happy she became because of the four hundred multiplier effect of the group. I wondered how the world will feel with nine thousand people meditating together in groups. It is like the science of photons contained together creates exponentially greater heat or more magneto particles create greater magnetic field. The AS_U particles enlivened in the human mind at the level of silence creates the Unified Field of bliss exponentially increasing with large groups together.

"Mind, body, behavior and environment improves tremendously with meditation," I said. "Never miss your meditation," I told her. There were some parents who came to see me. Some were extremely happy their children had made such a dramatic change in their grades and became surprisingly happier and well behaved.

A handful of parents were nasty, questioning what religion I was teaching the students. "Is it some kind of a new religion?" I was interrogated. "You do not play around with some kind of strange Hindu or Muslim religion with our children," they threatened me. "We are Catholics, you know." All the students were mostly Catholic and seldom are there Muslims in downtown Manila. They mostly live in the southern island of the country.

"Religion and science again, they are conflicting. It is the same cause for world conflicts, Muslims and Christians, how pathetic, as if one god of one religion is better than the other gods. As if religion can create peace or prosperity, on the contrary it is the cause of many wars. There are more poor people in countries controlled by organized religion," I thought to myself.

To dispel the conflict, I requested a Catholic priest, Fr. Nepomuceno, a friend to officiate Holy Mass every first Friday of the month. He was happy for the small donation I gave him each occasion. It became a regular first Friday school affair, like my TM school became an all Catholic school! The rumblings about religion faded away. There was a typical Catholic mass at eleven o'clock before lunch and then a lunch buffet prepared in the school canteen would follow.

At one o'clock in the afternoon, there were scheduled student presentations. A wooden platform was made covered with a nice-bright colored cloth where the students sang songs, American Idol-style. Songs and dances were presented not only from the students but from anyone including the cooks in the school canteen. It was such fun! Not every participant had a singing voice, but they contributed rhythmic sounds sung in a carefree, couldn't-care-less complete abandon, that created so much entertainment you couldn't stop laughing watching them joyfully perform.

There was so much evidence of happiness radiating from everyone. Trophies and awards completed the afternoon show. In the evening, at seven o'clock after dinner, the dancing would start. We hired a mobile disco for the dancing part. It was complete with blinking lights dancing with the music. It was such a sight! I remember the students and their guests dancing their hearts out, laughing and giggling in complete abandon all dressed at their best.

My daughters came with their friends too. I tried to dance the night away right along with the students and my children. I never had a chance to dance joyfully and enjoy parties like this.

In my youth, I was just too busy studying and working my way through college. Disco dancing was not part of it. Now, I was enjoying dancing like a teenager together with the students, giggling and laughing together with them. The happiness was just awesome!

You couldn't mistake the joy. It was magnified four hundred times inside your being. The joy was so intense and deep you cannot describe what was happening to everyone. They were just all giggling and exchanging happy jokes, in contrast to before learning TM, student come looking somber and curious eyed.

Every student looked forward to every first Friday affair and noted the event. All the students anticipated the exponential blast of bliss occurring each time. The student had presentations that were truly fascinating. We have the most enjoyable day every month. The fun was exponentially increased by the four hundred students meditating together.

This is how you feel when heaven descend on earth! Not only were we blasted with bliss, the surroundings also participated with joy. There was an exponentially higher level of happiness and fulfillment in all respects; individual, community and country. The value of real state in the area surprisingly boomed up and new buildings around the school area rose as the economy improved! Crime was reduced in the city! It was such a experience of bliss you could not forget. You just get addicted to it! It was the most memorable and happiest moments of my life, and until now, having grown old, I still cherish the happy memories and wish I can start a school again!

Nine thousand is needed for the world, a large group of people meditating together that is equal to one percent or the square root of one percent of the population of the world practicing the advanced TM-Sidhi program that Maharishi formulated. The world will be feeling an incomparable bliss like we felt then! Not one of us ever experienced that exhilarating level of bliss ever before.

Then in the southern part of the country, in southern Muslim Mindanao, where constant rebellions occur, the rebels laid down their arms. It was published in the news all over the country. The whole country appeared to have regained unity and stability after decades of rebellion and political struggles. It was just four hundred students in one place. Maharishi's goal is nine thousand in one place. Imagine the energy field effect that a nine thousand group will produce! Increase of bliss exponentially 9,000 times! One of the biggest groups I had the opportunity to attend was in Washington D.C. in 1993. It had four thousand participants! The result was exponentially more blissful. Crime rates went down. The state gained more stability. There were less traffic accidents and less health hazards. There was increased positivity in the area. There was bliss reverberating around and lingering in all the 4,000 participants, even affecting people walking in the streets of Washington, D.C. People in the streets were happily greeting one another.

"We have to create a bigger group permanently to create heaven for the whole planet," one mediator walking with us said as we crossed a street in Washington, D.C. The bliss effect was so enlivening. You just feel so wonderful. Everyone must be wondering where all the bliss came from. It comes from the natural phenomenon that occurs as soon as the group starts to meditate and quiet their minds together. We witnessed two years of growth within the city until it was closed. Two years of blissful existence stayed in my mind long after! We could not expand the school, as the area was not large enough for increasing more student population. The Department of Education advised as to look for a bigger area if we planned to expand. We tried looking around for a bigger area to expand. Then I heard about a property of the local TM movement that was vacant for many years. I told Maharishi the idea of a school there. The possibility of the seven thousand there was almost possible as there was not a school existing in the highly populated area. Downtown Manila colleges were more than two hours away by car travel excluding encountering the long unpopular Manila traffic. Maharishi said, “Go and put a school there.” We made hurried preparation to expand in Antipolo, Rizal province, the property Maharishi referred to. I hurriedly prepared for the school there following Maharishi’s instructions. However, an unfortunate event blocked the success of the project. It is the same conflict and power play that can occur in any typical group endeavor, one party and another party’s interest conflicting against the interest and benefit of the whole. I recalled back the grim memory of my own failed City of Happiness project in California for the seven thousand that I attempted to do. I had to console and explain to myself that the project was much too great an effort for one person like me to do. The seven thousand was formed in 1983 described in the Global Taste of Utopia in Fairfield. Iowa and was organized for barely twenty one days. Then in 1983, the assembly was supported by several American millionaires and scientists and the event lasted only for three weeks.

There we were, envisioning a permanent group of seven thousand with the energy and drive we could not imagine we could do. We worked day and night with a desire so great that night seem to be day and days were so full of energy that seemed to stay with us twenty-four hours. It was unfortunate, in one single turn of a key, the door was padlocked and my third TM school was closed.

Development of Intelligence
Increase IQ in University Students
Through Transcendental Meditation

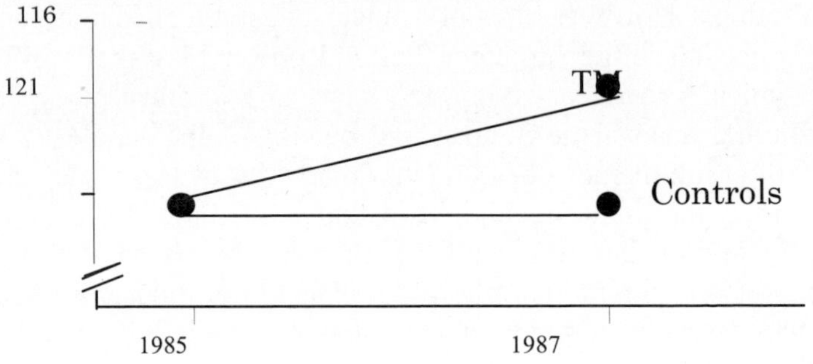

University students who were practicing Transcendental Meditation and TM-Sidhi programme increased significantly on intelligence in contrast to control students. ***Reference****: Personality and Individual Differences 12: 1105-1116, 1991*

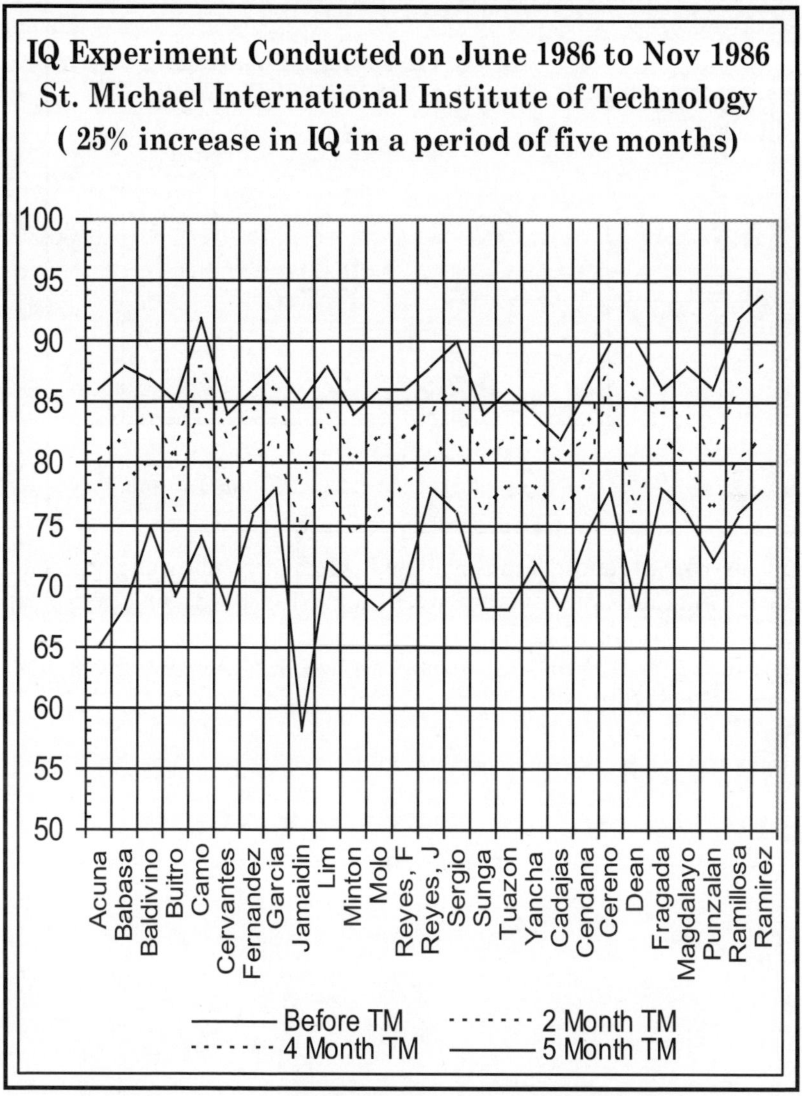

INCREASE IQ AMOMG STUDENTS PRACTISING TM

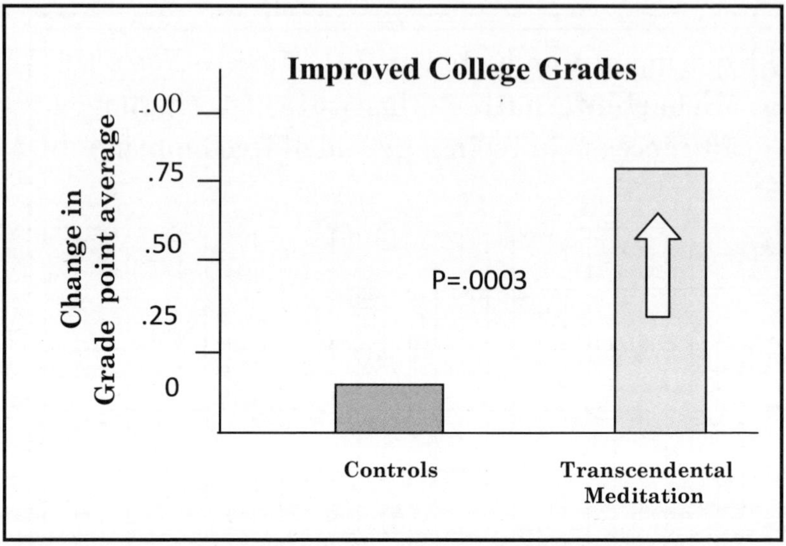

Students who practiced Transcendental Meditation showed significant increase in grade point average over the next three grading periods in comparison to control students.
Reference: Research on Transcendental Meditation programme, Volume 1:396-399, 1977

12

CLIMATE CHANGE AND FLOOD CONTROL

Modern technology has yet to come up with balancing climate temperatures in the atmosphere besides the reduction of greenhouse emissions which has been on-going for several decades. Dates of implementation for the reduction of greenhouse gases varies from country to country and many are still in transit. Cities and countries propose commitments from years 2100, 2030 and beyond, meantime, many flash floods and massive droughts constantly occur on earth.

The laws of thermodynamics express the behavior of air particles by variances in temperatures. The rotation of the earth around the sun determines the season, in summer the earth tilts nearest the sun, so the season is warm. In winter the earth is farthest from the sun so the season is cold. Just like at night, the sun is out on the other side of the earth, the temperature is cold or during the day, when the earth faces directly the sun, the day is warmer than at night.

Climate temperatures determine the equilibrium in balancing the behavior or properties of air particles in the atmosphere, like the behavior of air with heater on the inside of a car and cold air outside the car or vice versa, warm air cools and turns to water droplets, condensation is observed in the glass panes.

When air equilibrium sways in the atmosphere, rain and wild winds start to form. This is where climate change becomes destructive. When the earth starts to tilt further away from the sun, the greenhouse layer made of CO_2 starts to cool, it sucks the warm air from the sea and creates turbulence in the air, then wild winds occurs when air from the sea starts to cool and turns to water as rain. It is the natural character of air when cooled, it turns to liquid water, when further cooled, it turns to solid ice.

The rapid sways of hot air and cold air cause wild typhoons and hurricanes to build up causing torrential rains to fall and flash floods drawing winds that create high waves, fifty feet high or more flooding shore lines. Larger concentration of CO_2 in the atmosphere cause larger suction pressure of air furthering greater winds occurring causing larger torrents of rain to fall and flood the earth.

Hurricanes and typhoons seldom occur in desert communities where there are no large bordering waters, in contrast to communities surrounded by sea. The evaporation of sea water correspondingly cools and turns to liquid rain. The warm sea water is sucked by the cooling CO_2 layer in the atmosphere as low temperatures sets in creating wild winds and accelerates the cooling of the air and falling as torrential rains. Floods and hurricanes are observed to occur more often and in greater intensity creating massive destruction like that in the Joaquin hurricane during the first days of October 2015. It occurred in the Bahamas and considered to be the worst disaster in the shipping industry. It sunk the El Faro, a large cargo ship twice the size of a foot ball field with thirty three crew men on board who all drowned at sea. Then, it ruthlessly swiped South Carolina and it was described to be a flood happening in a thousand years, it caused massive destructions. Nineteen people were found dead. Many roads were eroded and dams destroyed with flood waters. The recent hurricane Joaquin signals that climate change is truly occurring. It followed several previous floods and hurricanes like that which made a touchdown in New Jersey shore in 2012 or the massive flash floods that occurred in Colorado, Texas or Florida in the USA. During Christmas time, before the onslaught of the New Year 2016, towns along the Mississippi river like Illinois and Missouri went under water, nearly thirteen feet deep flooding highways and sunk many homes.

Understanding climate change and the greenhouse gas effect unfolds an urgent need to find a solution to flood control. With the magnitude of harm it brings and the large costs to rebuild communities affected by floods, a solution to solve it should be considered by governments no matter how unconventional it may seem. Finding a means to establish equilibrium of hot air and cold air in the atmosphere relating to CO_2 concentration has never been before found. Like in other problems, alternate solutions are always explored. No one problem could have only one solution, there are always more options and sometimes they can come easy and cheaper.

Some climate change skeptics just shrug it off as just some same stuff happening in the long history of the planet. In the Middle Ages, lightning and air turbulence was considered to be some sign of an angry god and humans on earth has been at his mercy for many centuries. The problems happen more often and of greater intensity than before the Industrial Revolution that started in the 1750, there must be a reason why it is so. It cannot be anymore caused by a god angry all the time when floods, hurricanes and droughts occur.

Scientist found that the greenhouse gases made mostly of CO_2 cause climate change. In summer, it cause fires and drought, in the coming of the cold weather, the concentration of CO_2 cause fierce hurricanes and flash floods. The greater its concentration, the larger hurricanes, typhoons and massive floods occur.

Understanding the behavior of air particles in the atmosphere with its variations in temperature is just the beginning of understanding the solution. If there is more cold air in the atmosphere, it sucks warm air or vice versa. If more warm air is formed, it sucks cold air and the process results in air turbulence resulting in wild winds, torrential rains and flash floods. The concentrations of CO_2 in the atmosphere just repeat the same cycle of warming and cooling in the earth's atmosphere, the denser they are formed, the greater devastations on earth.

So the solution must be how to cool the warm air in summer or warm the cool air in the cold weather. The CO_2 concentration in the atmosphere sways extreme climate conditions on earth, if warmed cause massive droughts, if cooled it cause rains to fall bringing about flash floods of unprecedented proportions along sea shore communities. The reduction and elimination of CO_2 is a complex and tedious process and implementations has been extended for decades and on a long term basis. Its remediation affects economies of nations.

The agenda of climate change has been in place for many years, many countries presented their intents to reduce carbon dioxide and car manufacturers try technologies to utilize emission efficient cars, but it appears little is being done in an immediate manner. Droughts and flash floods continue to occur in greater number and of greater intensity.

Maharishi's discovery of the Unified Field provides significant solution to climates change and balancing air equilibrium in the atmosphere. It holds natural law's temperature mechanism of warming when the air is cold and cooling when the air is warm in the atmosphere. The Unified Field occurring within the human mind activates the natural temperature control mechanism of the human body in large groups together at the level of silence within the human mind, when warmed, the body activates its sweats glands that triggers the cooling mechanism of the human body. The human mind has the capability to balance temperature equilibrium in the atmosphere. Experimental evidences already exists.

The Unified Field created by large groups together at the level of silence provides an invincible shield from the cold atmosphere above and from the warm air evaporating from the sea. Warm air cannot penetrate the Unified Field created on earth by large groups practicing TM and Yogic flying together in one place. As soon as the threshold of the number required to activate natural law's character of invincibility, temperature equilibrium in the atmosphere is instantly maintained preventing wild winds and torrential rains to occur. Wind turbulence start to fade away. Climate change established by the natural temperature mechanism within the human mind at the level of silence starts to control the occurrence of flash floods and droughts on earth.

This theory of balancing climates and flood control was validated during the building of the thirty storey tower I attempted to build in 1994. As soon as four hundred workers on the job started to practice TM together, floods that regularly occurred near the area, just about a block away from the tower, it subsided and flooding completely stopped. There were no more floods happening, in contrast to regularly happening before the workers started the practice of TM together. The temperature mechanism within the silent state of the human mind in large groups together balance the equilibrium of cold and warm air in the atmosphere. When too cold, makes it warm, when too warm, makes it cooler.

I remembered before I started to build the tower with 400 workers on the job, torrential rains flooded the nearby streets. One time, after school, the van carrying my teenage children cannot cross over the flooded streets just a few miles from home. They were caught in the rain and had to sleep on the van parked on a side street along with many other cars stranded by the floods. There was no way to pick them up as the whole area was entirely flooded.

Luckily, the rain stopped and the floods subsided, early in the morning next day, the driver slowly inched his way away from the flood waters and slowly drove my children safely home. My youngest daughter was crying and chided me saying, "Mommy, you said Maharishi can stop the floods. What are you doing? It is good we had a security guard with us, otherwise we would have been so scared sleeping in the streets under a wild storm. We were lucky the rains stopped otherwise we would all be drowned in the floods."

My daughter's message remained in my heart that never went away. Each time a storm occurs, I remember the fear of every mother for the safety of her children. I am overwhelmed each time head counts about how many drowned in the floods. Every war and conflict occurring in the world, I think of the mothers and their children. I grieved for the mother and her two children who drowned at sea fleeing war torn Syria. I validated Maharishi's programs eliminating suffering on earth because I want my children to be safe.

I was not the only one who validated Maharishi's principle of invincibility. It was validated fifty times and displayed the same results, climates balance. Peace and prosperity prevail.

I bow down to Maharishi, who has discovered the natural law principle of invincibility that can remove all suffering of humanity since time immemorial. No one in the long history of mankind has brought about the Unified Field that can easily transform and eliminate miseries of human life on earth. To me, Maharishi is the greatest scientist surpassing any ancient master for bringing about a technology that can remove suffering and can be easily implemented with least cost.

It took sixteen hundred years before Galileo's theory of round earth was accepted, reversing Ptolemy's flat earth theory. Maharishi's theory was accepted by two governments during the past decades, the government of Mozambique during former President Joaquim Chisano who formed large groups of military practice TM and Yogic flying and in the same year, Cambodia's government in 1994 provided a land where a school was built where students practiced Transcendental Meditation together. The group practice of TM removed negativities to occur, as Maharishi has foretold.

It is therefore the responsibility of every government to test and validate the Maharishi Effect and create a permanent group of TM and TM-Sidhi program that Maharishi has brought to light. Instead of the many floods and droughts regularly occurring, an experimental model can be made in every country or the 9,000 in one place on earth on a permanent basis. It will guarantee happiness and fulfillment of every individual, community and nation on the planet earth.

13

ANOTHER TM SCHOOL CHANGED THE CLIMATE

My second TM school had to transfer to a larger place as the land was not large enough for added student population. A property was referred to me that used to be the place where Maharishi first started a group in the Philippines. It was there that Maharishi formed a group of TM and Yogic flying in 1979 about two hours east of Manila. They were manufacturing rattan furniture at that time from Monday to Friday. Maharishi called them back to work one weekend as a strong storm was scheduled to touch down in Manila that same weekend. As Maharishi has predicted, the strong storm made a turnaround direction away from Manila as soon as the workers resumed their regular TM and Yogic program twice a day, before and after work to the astonishment of the Philippines weather agency monitoring its path. Soon after, the building they used was abandoned.

I spoke to Maharishi about expanding my second TM school there, instead of being vacant for more than ten years. I told Maharishi, "There is not a single college in the area with nearly a million population. We can probably get 7,000 students there." Maharishi said, "Go and build a school there." The prospect of the 7,000 students appeared easy as the colleges in Metro Manila were two hours away and with the on-going heavy traffic along the main highway, travel time sometimes doubles up.

We hurriedly disposed the rattan furniture stored there and partitioned the building into school rooms and my staff monitored that all the utilities were replaced and upgraded. The area was large for a start, about nine acres of campus with some coconut trees scattered along the sprawling landscape. The prospect of the 7,000 provided great inspiration for us all and made us all extremely happy diligently working to finish the facilities ready for the opening just a few weeks away. We were all driven to fulfill the goal of Maharishi in 1988 to form the 7,000.

We followed the same process we did in previous schools, distributing flyers in churches and public places stating that a new school was ready to open. On the first day of enrollment, many students were falling in line to register, but on enrollment day, I was surprised to find the main gate padlocked. It was a tragic incident, for personal interest and for probably a meager commission amount, it was sold big time because of the renovations I have done, instead of its dilapidated state that remained unsold for more than a decade. The third TM school closed before it even started.

Many of my staff cried with me. "That crab and scorpion conspiracy sold you out," they cried referring to the birth signs of the conspirators. It is the same cause of conflicts constantly happening among parties, one party overcomed with greed for just a few pieces of gold for themselves, while the other party, instead, do what benefits the majority in the population.

I was broken hearted, not because of the large amount that I had spent in rehabilitating the place, but for the lost opportunity of establishing the 7,000 Maharishi that wanted formed for so long and tried to form many times and for so many years that was needed for the transformation of the world.

A large coconut fruit fell direct on the head of the father of the woman who padlocked the school and died that same day. Someone said, "Natural law does not allow anyone to block the pursuit of nature's cause. Natural law was angry with the family of the woman who padlocked the school, the coconut killed her father who was standing beneath the coconut tree watching the crowd of students falling in line to enroll." Strange, the same thing happened in my previous school. The student who was leading the rally against the teaching of TM in my school, he slammed the school gates on me nearly smashing my hand. On his way out in the highway, his car suddenly rolled over and he died the same day. The same happened in my "City of Happiness" project in California, also a project for the 7,000. The man who instructed a killer bee hive be planted in my yard to scare me away from the city, slipped near my garage, broke his neck and soon died. Then, I remembered the same statement that one of my staff has said, "Natural law does not allow anyone to block the pursuit of the cause of natural law."

The 7,000 project was an endeavor probably too big for one person like me to undertake. Maharishi attempted to do the same for many years with many wealthy Americans and other patrons around the world who donated money, but to date the 7,000 has not been formed. About 1,000 are currently at the Brahmasthan of India and similarly are in Fairfield, Iowa, USA.

Later, I tried again, it was the fourth time I established a school where students practiced Transcendental Meditation. Luckily, the land of the second TM school increased in value a hundred fold when the thirty storey tower was build on it, then, I recovered the money I lost in the third school that was padlocked and closed.

The fourth TM school came about when a friend invited me for a weekend by the beach, she said, "You are always working. Let's go to the beach and stay there overnight." She said, "I was invited by the Mayor of the town, in Bolinao, six hours drive from downtown Manila. You and I will be his guest." At dinner, the Mayor said, "We need a school here as the colleges are so far from town. Students have to travel over two hours. There is land here that you can buy and build a college." On our way home the next day, we checked out the property which was nestled right on the main highway. I snapped it up, bought it right then and there. Then I directed the architects and engineers of my company to work and prepare the plans for a school building. Twenty large class rooms on a one storey building soon rose on the sprawling Bolinao town campus. Like the previous schools, we had to get out the word of a new school opening in the area. We prepared flyers and distributed them in churches and public places. We managed to gather five hundred students again to enroll in the newly built school.

One time, Maharishi urged me to take charge and fight in the frontline, wherever that was or what he meant then. He wanted me to refer to it as my school. In the past, I always remained in the background. Maharishi instructed me, "Your decision will always be correct. Do, what you need to do," he said very strongly. His instruction was urgent, "Do it. Just do it," Maharishi said to me.

My eldest daughter said the same. "Just do it, Mommy. You are the better one. You buy the land, build the schools, and pay all the salaries. You even pay for the TM fees." This was my own daughter complaining about some uncooperative local people.

"And at one time you paid the salaries of the people just to meditate who did nothing but complain. Even Maharishi, he just sits and talks and talks. You go around working and working even on Sundays and holidays."

I reminded my daughter, "When you work for the cause of natural law, there is more opportunity to make money than when you are working for yourself. When you work for the cause of TM and natural law, natural law supports you. I have experienced this. But of course you should always have a group around you, if you are alone, it's different. The exponential effect operates in groups. That is the secret of Maharishi's teaching so few understand. Did you ever want something, anything that you did not get, or travel anywhere you did not get to visit?" I cooled off her fretting.

"I just wish some people in your TM movement help you, Mommy. No one does, just you. I have people here with me helping," referring to my staff. "The credit is yours." Maharishi once said. I answered back my fretting daughter.

"Then, I alone earn and take credit, not them. There will always be people talking, doing nothing. They do not understand Maharishi's teaching." I said, "When you work for TM and natural law, you become master of natural law. Maharishi says this many times." It is like nature saying, "What is it that you want, Master?" Nature becomes your genie, like the tale of Alladin and the genie in the magic lamp. The genie creates the magic, not Alladin, the master. Those that serve are those who get support of nature not the masterminds that do nothing but sit and talk."

The students started to meditate in mid-July. The month of June was the same scheduled school opening in the country. I always scheduled teaching TM on my birthday in July, one month after school opening, like a reference point for observing the effects of TM to happen. The day was so warm and humid. Just like in my previous TM school, rain fell on the roof of the school the moment the students closed their eyes to meditate and the rain stopped after the student opened their eyes. There were no dark clouds that foretold the rain. The same event in my previous school happened the following July and on my birthday too. The coincidences were just too awesome to recall. It was the natural occurrence that Maharishi referred to. This time another TM school balanced the hot July climate. Like the rain that fell in Brazil during the Taste of Utopia. The same occurred in Mozambique. Rain fell in the drought areas. The same natural phenomena occur as the minds quiet down during TM. It became a fact to me after several tries and similar effects of climate changing!

I was just so excited with the idea of a new school. I bought the land and built the school unaware that it was a rebel area. The military called it a red alert area. "Of all places, why did you buy the land there? It is one of the poorest towns in the Philippines," a friend said. I was totally unaware of the peace and order situation in the town of Bolinao. The town was more than six hours north of Manila and we never heard or read any news from the remote poor town of Bolinao.

The benefits are too great to miss it for anything. It was enough for me to know there was no school in the area. And to make the decision seem more secure in my mind, it was the town Mayor's idea to put the school there. I concluded then that it would be a safe place. But in my first visit to the construction site, I was told that ten armed men robbed the worker's food storage adjacent to the place where the workers eat their meals. We brought only one security guard, Gerry, who was the crane operator in the tower that we were building at the same time, later he became one of my trusted security guards. He carried only a small hand gun.

He warned me, "Two weeks ago, some fifteen military personnel were ambushed and killed about ten miles from here. No one wanted to tell you about it because you might get scared coming here to the construction site." The massacre was truly scary and frightened my staff and me too. We had no option to back out after finding out the rebels were near the school we were building. We cannot stop the construction of the school building that had already begun. I was undaunted. I believed in Maharishi's technology of defense, After I heard the news that the rebels were near the school area, I immediately instructed all the construction workers and staff to learn TM. It was my only option. I had no army to protect my staff who were working in the school from any harm the rebels can inflict to them.

One by one, they all learned Transcendental Meditation. The most important responsibility of the supervising engineer was to have the workers practice TM twice a day, before and after work. It was the first report the project engineer had to provide me every day. There was a small hut in the construction site where my staff and I stayed overnight, but to avoid any undetermined risk. I was advised not to sleep over in the construction site anymore.

"This is a dangerous area for you, this is a rebel area." Greg, the project engineer said. Greg was the senior engineer we called for construction work. After the rebel scare, I visited the school site only twice a week, round trip, back and forth alternating cars, the gray Nissan car, then the red maroon van we fondly called the 7000 express. We would leave downtown Manila at five in the morning and arrive in Bolinao before lunch.

I am up usually early ahead of my staff, watching them scamper hurriedly to join me. It was such a funny scene seeing them running barefoot to the van carrying their shoes, sometimes their hair still dripping wet from coming out of the shower room, dragging with them their belongings. It was peculiar seeing all of them so happy and joking at five in the morning. Meditation makes them happy at any time of the day. I had the same strange happy feeling too. We would all meditate in the car except for our driver named Gingo, who really liked to drive fast, full speed ahead, he would meditate upon our arrival in the school site.

I chided Gingo one time, "Sometimes, I half-expect you to pull back on the wheel and take this car to fly. You drive faster than a plane before it leaves the ground." There were resounding laughter from everyone. I couldn't help remarking what a happy group we were. "The workers are happy too," Greg, the project engineer said. "No one complains even if they have to work more hours to meet the school opening. They all meditate, that's why." Greg supervised the construction and the drilling of the deep well as well. He was never a school administrator. His job was in construction.

After a hectic month's schedule, back and forth in one day from Manila to the town of Bolinao, I was hospitalized for a week due to fatigue. Twelve hours of travel back and forth was too much for my physiology. I guess even if the mind is willing, the body was failing faintly.

"Nothing wrong," the doctor said. "You just need rest. Take these vitamins to regain your strength, and never travel back and forth, twelve hours in one day again." We installed a mobile phone at the site, but the reception was bad.

Then, there was no mobile server in the remote area. We had to drive back and forth again, but not in the same day. We stayed in an adjacent hotel along the highway near the waterfront at night and left for Manila the next day. It worked well. In three months, even with the tight construction schedule, the school was nicely completed. We had all the faculty chosen and hired. The library was filled with books. The computer room was humming with computers and ready to use. I hired the town Mayor to be President of the school.

Enrollment day came, Monday. No student appeared. Tuesday, Wednesday, Thursday, no student came for enrollment. We were all heart-broken. I called everyone for a staff meeting. "What happened? Did you distribute the flyers to the different high schools in the areas as I instructed you to do?"

One of my secretary, Bebeth raised her hand and said, "The flyers arrived after all the schools closed for summer. You were out of the country, in California then. We were not able to distribute them. They are still here." I realized, we were so excited about the school, that we miscalculated the time table to distribute the flyers.

"What do we do now?" I asked. "We didn't go to all this effort, buy the land in a dangerous rebel territory, quickly construct this beautiful building, and spend so much money, time and effort for nothing!" I was pacing back and forth in the room, very desperate! Not again, I thought, remembering the school that was padlocked with students falling in line to enroll! This time, this school gate was wide open, the grounds so nicely paved and landscaped, but no student falling in line! There was no one, zero! Sonny, one of our senior staff, stood up and said, "There must be something we can do."

"Okay," I said, reading his mind, "be quiet and let us meditate." The silence was so deep and eerie. A few minutes passed. After twenty minutes of meditation, Sonny stood up and said, "I have no other idea but we should all go to church on Sunday, tomorrow," he said. His reaction was typical for someone with a problem, go to church and pray and hope that God will provide the answer.

"Yes, yes," I said. "We go to church today, now, not tomorrow!" Everyone was surprised and stared at each other wide-eyed in disbelief. "Divide yourself in groups of ten and bring a letter to the ten parish priests of the ten towns surrounding the school. Sonny, type this letter," I instructed.

The letter was hand-carried to the ten parish priests that same Saturday afternoon. Enclosed in each letter was a few hundred pesos donation handed to the parish priest. Sunday was the big day. In the Sunday sermon, the priest announced that scholarship was being offered by the owner of the new school. It is available for anyone who will enroll anytime starting tomorrow. Next day was a surprise indeed! One hundred fifty students came the following Monday. A week later, another one hundred fifty students came to enlist. Everyone was excited. We had three hundred students enrolled in two weeks!

"We need two hundred more. Our target is five hundred," I said. Once again, a staff meeting followed by another twenty minutes of deep silent meditation. "Tell them, the ten mayors and ten barangay captains, to come and bring ten scholars each. Twenty persons bringing ten scholars each will bring us two hundred more," I instructed my staff.

The letters were delivered with a few hundred pesos donation. There were ten towns around the school with ten mayors and ten town barangay captains, who are mayor's assistants in the different areas of the town.

On the first day of school, our target of five hundred was reached - exactly five hundred! You cannot imagine the happiness and excitement from the entire staff. It was like a miracle happening! It wasn't a miracle, but a natural phenomenon. It was the result when groups practice Transcendental Meditation in groups and desire together, desires manifest spontaneously!

It was the most significant milestone that made my belief in the Maharishi Effect unshakable. We desired five hundred in groups together and exactly the five hundred came so easily without effort! Meditating together in large groups get desires fulfilled easily! It is the science of the Maharishi Effect in action! Maharishi's Yagya program uses groups desiring together now being done in India where nearly 1,000 pundits meditate together.

When the electrical bulb reaches the required wattage, the bulb illuminates. We used the same energy like electricity! Switch on the group desiring five hundred, the result is a group of five hundred! The "sap" lingering in the human mind got enlivened and created the energy that produced the five hundred students that came to enroll.

Thought force at the silent level of the human mind in groups together contains an energy field described in the recent discovery of the Unified Field[38] by modern science that has its basis in the ancient Vedic science where Transcendental Meditation originated. Consciousness contains that science of the mind that creates fulfillment of desire. The mind has to be brought down to the level of least excitation, at the quietest level of the mind. Transcendental Meditation is a simple procedure that quiets the mind.

I remembered Maharishi's quote, "It is the purity of a man's heart and mind and his innocent and faithful approach to action with the purpose of all good to everyone, which really succeeds in yielding maximum results with minimum effort." It is magnified by group minds desiring together for a common cause improving the general well-being of man and his surroundings.

I will never forget the inaugural day of the school. I remember how beautiful the school was! There were flowers all over the pathways and on the corridors on all sides. Flowers spread all over the building. It was like a new garden that quickly sprung and was blooming all over! There were so many pots of flowers, I had never seen so many flowers in my life. Flowers of every color, in pink flower pots placed in every corner! It was indeed a heavenly sight. The girls wore white uniforms with pink ties and the boys, white shirts and khaki pants.

Two Maharishi's Purusha, Maharishi's select male group from Fairfield, Iowa, USA, David and Stephen came for the inauguration. Stephen said, "This town will be blessed." Every community with large groups meditating together grow in peace and flourish, in contrast without a group, crimes constantly occur and there is poverty. That is why Maharishi's urgent instructions was to establish large TM groups in every community or country to establish

[38] *See Chapter 8, The Discovery of the Unified Field*

prosperity and prevent crimes. Many shootings occur in the USA. Many terror groups has risen in the Middle east! They need to establish large groups practicing TM and Yogic flying together.

During the school inauguration, five hundred students were all seated behind the many guests in the front of the audience. I felt so much excitement and eagerly anticipating the awesome effect that will be produced as the students started meditation.

I was introduced and asked to speak. The same message that Maharishi repeated in his lifetime, I repeated it once again. "At the silent level of the mind…."

As soon as the students started to meditate, it rained for twenty minutes. The same phenomenon happened with the students meditating in the schools that I previously built. The same rain phenomenon occurred in Brazil during the Taste of Utopia experiment. The same happened in the drought areas in Mozambique, Africa. The group effect balanced the warm climate.

There were rebel groups in the area. The town Mayor had not told me that Bolinao was a dangerous rebel territory. After meditation started, the rebels laid down their arms as far as the Muslim southern part of the Philippines. It was all written in the newspapers!

The rebels just vanished from the Bolinao town as soon as the students began to meditate. Rebels are born because of the failure of governments to provide a better living environment for its people. Rebellion occurs because of the frustrations of the people and their inability to fulfill their needs. Rebellion is not the solution to poverty. Poverty is the common cause of war and people protests.

Then later, after the school was built, people in the town came to the basketball court that was built for games and social programs. There was also a community disco dancing that the school scheduled in the area. The same happiness radiated so conspicuously from everyone, quest and students were blasted with bliss! The bliss particle found to linger within the human mind became alive and everyone was bubbling in happiness so noticeable from everyone.

People became friendlier and happy. Not like before the school started meditation. People came to the school campus suspicious and curious about what kind of school we were building. There was peace, not only in the Bolinao town, but the whole country. The country's economy began to improve. There was a real estate boom. Floods and typhoons that were a common occurrence subsided.

I was observing the events everyday with great excitement and I was truly humbled by our accomplishment. I brought some staff with me one day to make rounds around the nearby towns. We saw some dramatic changes happening in town. The magic of the Maharishi effect had begun! We saw the unfinished and abandoned bridge connecting Anda Island, a small fishing village adjacent to Bolinao town being renewed and under construction. It was a bridge that had previously been abandoned for more than fifteen years. Many town onlookers were curiously watching the rusted old bridge being removed and replaced. With so much awe and excitement, we all watched the bridge being completed! The fishing island was a main source of fish for the people in Bolinao town, soon fish became abundant and easily available when the bridge opened.

The market place that previously was a sidewalk flea market and a spot store, where temporary vendors just took a spot and sold their wares around the street, it completely disappeared. The open sidewalk flea market was gone! A new market made of a two-story sprawling concrete building was recently built!

The unpaved highway and roads leading to the town of Bolinao were finally paved with concrete instead of the previous dusty gravel roads. In another nearby town, we saw another market being built too, with new concrete facility, not anymore like the open sidewalk spot market seen in every poor remote towns. Nearby surrounding market scenes changed soon as the student started to meditate. All the old scenes spontaneously changed at the same time. They were unbelievable coincidences.

I felt like I was doing my daily rounds as a project engineer, only it was the Maharishi Effect project unbelievably sprouting up simultaneously in the sleepy remote poor town of Bolinao and beyond. The impoverished towns surrounding the school were enjoying the magic of the Maharishi Effect too. The magic came from the beautiful pink colored school building recently built, the Natural Law Productivity College.

Some magic happened along the highway as we were going back home to Manila from the school in Bolinao. Our pickup truck driving ahead of the van we were riding on was hit by a large passenger bus filled with people as it turned on a curve.

The truck turned around three hundred sixty degrees right in front of our very eyes, just a few feet ahead of the separate van we rode on. It careened down the road curb and stopped down the bushes nearby. Gerry, the driver was stunned, saying, "Am I still alive?" He was not a bit hurt. The pickup truck showed no damage either. There was only a small dent near the left front wheel of the truck. We were all surprised no one was injured. The bus that swiped the pickup truck stopped and all the passengers clapped their hands upon finding out that no one was hurt. It was strange, our driver was aware he was turning the wheel 360 degrees like he was in a circus. It was like a magic stunt! It was unbelievable no one was hurt at all.

As expected, the group practice of TM in the school provided effects that showed the same positive results as previously accomplished in similar TM programs in other parts of the world. Like an airplane that overshot the runway in Malaysia, with two hundred passengers. No one was hurt. That was during the Global Taste of Utopia event in 1984! This one was Bolinao town utopia with only 500 students meditating together, but the results were the same! The Global Taste of Utopia experiment in Fairfield, USA, in Cambodia, in India and in Mozambique, they all showed similar observations. The periods of meditation dictated the same phenomenal positive effect like the small scale effect in our Bolinao school that I had recently built.

In the Miessner Effect in physics, no impurity can exists. It is the same science of particles in the Unified Field effect, no negativity can exist. The Maharishi Effect refers to the finest of all particles that displays natural law principle of invincibility that Maharishi espoused throughout his life. What was happening was not by chance or coincidence. It is the effect! It is the same effect every time there is a large TM group formed.

The economy in Bolinao town rose more than ten-fold to be able to build the new market, a sprawling two-storey concrete building instead of spot stores like swap meet market places. The income of the town increased and able to rebuilt the old steel bridge that was abandoned for fifteen years. The long dirt road highways leading to Bolinao was concrete -paved and must have been a hundred-fold increase in the town's income to afford miles and miles of new concrete paved highways. Prosperity increased in town! And all happened when 500 students meditated together in the new school that I had built in town!

More than fifty large experiments done at different locations around the world resulted in similar observations. Countless small TM events duplicated the results in small sale. The Maharishi Effect for the group practice of meditation is truly phenomenal.

Like take a bar of magnet, join it together, it will create a much bigger and exponentially greater magnetic field effect than just produced by one magnet. To date, the super radiance effect, produced by the group practice of Transcendental Meditation, surpassed any discovery of science relating to human well being.

Negativities in the environment disappear, climates balance, peace prevails and the economy improves. Small poor town prosper in an instant! All the positive results that Maharishi claimed would occur when TM and the advance TM-Sidhi program is practiced in large groups together in one place, positivity happens spontaneously.

A guest, who came to visit me, read a quote at the back of the second book that I had previously written about Maharishi being the greatest scientist. I gave my old book to him. On the back cover, I wrote, "To me, with all the outstanding results, repeatedly verified, to date, Maharishi is the greatest scientist that has ever lived. He has surpassed the teachings of all the ancient masters. He unfolded the seed form of all the laws of nature contained in the human mind. Such an easy simple procedure, nature designed it to produce such a massive benefit to the environment. No one in the history of the world has come up with a systematic and easy procedure to eliminate human negativities with one easy stroke."

I was very confident to make the statement, because I had actually verified it in my own small way, not once but several times. More than ever, after I read someone wrote the same, that Maharishi is "the foremost scientist in the field of consciousness and was considered to be the greatest[39] teacher in the world in our generation." I was not the only one saying that "Maharishi is the greatest!" It did not matter much to me what other people say, or what my guest said, "You must be Maharishi's pet for speaking so highly of him. The greatest...."

He nodded his head and slowly repeated, "The greatest," seemingly doubtful. He seemed perplexed; there must be some ambiguity about my quote. I assured him. "My quote is accurate. Do you know anyone or any system at all, in all the history of mankind that has improved society in one easy, single stroke? We are in the twentieth century, two thousand years after the death of Jesus Christ, the God of the Christians, we still have wars and terrorists exist around the world." I emphasized.

"Can you name a system like this? It seems that now, there are more diseases, incurable, like cancer and aids plaguing every nation. Children with autism appear to be on the rise. In the old days, there were herbalists who provide some form of drink to ease pain. In some ancient cultures, elders would chant over the sick person. Now hospitals are filled with many people falling in line long hours to be treated."

---

[39] *Appendix B, The Science of Being and the Art of Living, Maharishi*

"Many prisons are overcrowded with unavailable option to reduce crimes. There are fighting and wars happening, just like in the 9/11, in the USA, in Iraq, Afghanistan, Uganda, Darfur, Tibet, Africa. Bombings here and there, people protests in recent Egypt, Libya, Yemen Syria and Ukraine. Billions of dollars spent on wars. Many countries with falling economy, huge deficits, recession, earthquakes, floods...." I enumerated.

"More and more people are violating laws and prisons are filling up. Did you see the long lines in the public hospitals? You need to wait hours, even days. Did you know that? There are a lot of shootings in schools and… I stopped enumerating. He nodded his head and said overwhelmed, "Okay, okay, maybe you are right. Your Maharishi is the greatest, if he can improve all these."

I stood up to him and told him, "His Unified Field, the science of the mind can definitely do it. There have been many experiments and it has been proven correct time and again," he was overwhelmed by my statements.

Before Maharishi passed away in February 5, 2008, he founded the Global Country of World Peace, a country without borders and a home for peace loving citizens of the world, www.globalcountryofworldpeace.org. It details all of Maharishi's programs and named all the administrators in every country around the world under the capable leadership of his appointed successor, the world-renowned neuroscientists and medical doctor, Dr. Tony Nader, MD, PhD whom he honored with the title, Maharaja Adhiraj Rajaram and the responsibility of guiding the Global Country of World Peace.

Before he retired into silence, Maharishi announced, "I can only say: Long live the world in peace, happiness, prosperity and freedom from suffering."

"For all millennia to come, the world is going to be a happy world. The future of the world is bright, and that is my delight."

-Maharishi, January 11, 2008

I bow down at the lotus feet of my master, Maharishi for bringing about the supreme knowledge contained in the Veda and Vedic literatures describing the origin of Transcendental Meditation.

After Maharishi passed away on February 5, 2008, I covered a small table with a white cloth in my office and placed Maharishi's picture on top. I place fresh flowers every day to keep his memory alive in my heart.

14

NATURAL LAW PRINCIPLE OF INVINCIBILITY

Nature has many laws. There are the laws of motion, the laws of thermodynamics, laws of gravity, and now the recently discovered laws of quantum physics that has brought to light the character of the Unified Field that Maharishi referred to contain the seed form of natural law.

There are laws governing energy fields, like the field of electricity, magnetism, gravitational field and recently the discovered Unified Field surpasses any field discovered to date and holds natural law principle of invincibility. In the electrical field, it will burn anything that comes into its field. The stronger and bigger the field, the greater is their invincibility. The field they produce contain fine particles from the breaking down of atoms arising from matter as it turns to liquid, then further to air and gas described in the laws of thermodynamics.

Quantum physics described the character and attributes of the elementary particles of nature as they break down to finer particles from gas form to nuclear and finer forms. In the electric field, it contains the neutrons and protons particles. In the magnetic field, it contains the magneto particles which are finer forms of particles much finer than the proton and neutron particles. In gravity, it is the graviton particle that creates the gravitational field.

Scientist say, the finer they become the more exponential power in highest degree they create. The Maharishi Effect displays that power of the finest particles at the level of the Unified Field. The finest of them all, Maharishi called the "sap" that permeates the whole universe that creates from "itself, within itself and by itself.". I referred to it as the anima solo particle, the AS_U particle that animates the whole universe. The ancient Vedic literatures described the power at the level of "Brahma." The Maharishi Effect displayed the power Unified Field at the level of silence reached during the practice of Transcendental Meditation which has its origin in the Veda from the ancient Vedic civilization of India.

The magneto particles which produce the magnetic field that surrounds the earth is invincible to solar flares and protects the earth from the heat of the sun, otherwise no life form will survive on earth.

From the events illustrating the Maharishi Effect, the Unified Field effect is created when large groups of minds together contained in one place practicing the TM and advanced TM-Sidhi, or the Yogic flying creates invincibility from negativities, only positive trends in man's surroundings can exist as displayed in the Global Taste of Utopia event described in Chapter 6.

It is like in the Meissner effect in physics, when more magnets are joined together, they produce an invincible field, no impurities can penetrate the magnetic field they produce. It is unlike the power of the atoms in concentrated nuclear form of hydrogen particles that blasted off the whole city of Hiroshima in Japan ending World War ll. The AS_U particles are the finest and purest particles in nature, does not contain any negativity and does not cause destruction, instead it nourishes the whole environment. It is the particle of nature in seed form that pervades the whole universe. Some scientists call it the God particle or the particle of everything, where all matter and life forms in the universe spring from.

The knowledge than man posses this primal particle within the silent level of his mind, brings man to the most supreme status of man's destiny discovered in this generation. It was contained in the Veda from the ancient Vedic literatures, "I am Brahm, I am Totality," the Vedas said.

Man now can reclaim his birth right, and live an invincible life beyond suffering. He can exercise this great power to eliminate human suffering. He can now prevent destructions to prevail on earth. Climates will balance. Economies will rise. Wars will end. Man now can create heaven on earth. The Unified Field is a recent find that can do it. The process is easy and effortless, just by forming TM and Yogic flying participants in large groups together equal to the square root of one percent of the population of each country or the world, humanity will be transformed from the many maladies currently occurring in the world today to a problem-free world and everyone will be bubbling with deep bliss that can be drawn at the level of the Unified Field found within the silent level of the human mind. Maharishi said, "Like a tree, just water the root and the whole tree will flourish. The leaves will turn green, the flowers blossom and bears fruit." The Unified Field nourishes all life in creation.

Meantime, eliminating and preventing the harm done by climate change becomes an urgent endeavor by nations, like the super typhoon named Haiyan, that made nearly the most horrifying landfall on planet earth. The giant forming turbulence was seen from the International Space Station hovering thousands of miles above the clouds and above the earth. It identified the place and date of downfall, November 8, 2013 at the central island of the Philippines.

Locally named Yolanda, it was recorded to be nearly the strongest typhoon ever to have made a landfall on earth. Haiyan caused massive devastation from the far west to the east end of the whole central island of the Philippines with its astronomical circumference extending nearly 300 miles of condensed rapidly circulating wind power in a massive air particle turbulence ever to land in the country's history.

A landfall velocity of nearly 195 miles per hour swallowed everything along its path causing so large and massive destruction never seen before in the country. It completely wiped out communities and killed nearly 5,000 people and injured thousands in the population leaving a horrific scenario of shuttered homes and flattened structures. Thousands were thrown scattered inland among piles of debris and hundreds strewn away into the sea and drowned, leaving seven hundred thousand people homeless. Billions of properties and infrastructures were toppled down and completely destroyed.

The massive relief operations could not be led alone by the under-manned and limited capabilities of the Philippine government stunned with the unprecedented destruction so massive never seen before in its history. Foreign governments particularly the US government provided rescue teams and aid. Millions of dollars poured in to help the massively devastated country.

Many groups blame man-made carbon emissions for climate change. Others climate skeptics say it is not man-made. Until now no one has come up with a scientific solution to instantly tackle these massive wind particle disturbances so often occur bringing about a tragedy such as the super typhoon - Haiyan that made a destructive touchdown in the Philippines.

If climate change is not man-made, what made the air so destructive and powerful occurring today that never happened before 1750, the start of the Industrial Revolution? What strong wind force created that caused to completely devastate a whole community to scatter and disappear into the sea? The laws of thermodynamics describe air equilibrium as the balance of the same volume of hot air that confronts the same volume of cold air, equilibrium is maintained. In contrast to typhoon Haiyan, which must have held a large mass of cold air that dangerously sucked the warm air evaporating from the sea heated from the sun, creating destructive powerful winds of 195 miles per hour that made dramatic climate change into the most powerful typhoon yet to land on earth. With this display of wind velocity never happened before in the most recent history, there must have been a large layer of cold air in the atmosphere that caused strong winds to occur. It can be explained by the laws of thermodynamics. A great mass of cold air can create massive winds to cause the destructive rush of warm air evaporating above the oceans. In contrast, equal volume of warm air and cold air results in air balance and equilibrium, no turbulence can occur. The rush of warm air from the oceans sucked by the massive layer of cooling CO_2 in the atmosphere displays the drastic changing of climates, from cold to warm or vice versa, when warmer air from the sea to cooler air in the atmosphere due to the changing of seasons on earth. When cold weather starts to set in, that means when the earth starts to tilt away from the sun, beginning in early October until the coming of winter, it starts the confrontations of air equilibrium from to cold or warm or vice versa causing wild wind turbulence occur.

The strength of the rush of wind build ups are determined by the variances of temperatures and volumes of cold air or warm air confronting each other. The massive volume of cooling CO_2 gathered in the atmosphere cause the rush of wild winds from the evaporating air from the sea waters warmed by the sun.

I concluded that the earth tilts away from the sun when I noticed that the sun's shadow started to cover the terrace where I have my early morning breakfast. Later, I had to move further down a few feet north from the open terrace to the existing lawn to catch my morning sun which starts early in October.

As the earth tilts away further north, the concentrated CO_2 layer above the atmosphere starts to cool and sucks the warm evaporating air from the oceans heated from the sun, causing wild air turbulence producing massive winds as large as 195 miles per hour. It illustrated that the concentration of CO_2 existing in the atmosphere has dangerously increased, it caused such large wind power swirling in rapid speed during the Haiyan typhoon in the Philippines in November 8, 2013.

All the other hurricanes of lesser magnitude that made touchdowns in the USA like sixty tornadoes at the central part of the USA just a week after Haiyan in the Philippines, the Katrina in New Orleans, Sandy hurricane in New Jersey and the Colorado fires killing twenty two fire men followed by a massive flooding of biblical proportions caused so much horrific devastation that these natural occurrences, if there are solutions, however unconventional, need to be and has to be paid attention to. The invincible principle of natural law must be applied.

In March 2011, the huge tsunami in Japan stunned the whole world. Multitudes of people killed, homes, roads and infrastructures destroyed in Japan costing billions of dollars. Huge amounts of money were eaten up and destroyed by tsunami, typhoons, floods and fires. In the Haiyan typhoon, nearly one third of the area of the Philippines was devastated and billions of homes and infrastructures destroyed.

When the wrath of nature strikes, it does not discriminate how small or how big a nation is and neither does it choose how poor or how rich it is. Any country can be a victim one day or another in the future, with typhoons and hurricanes occurring more so in communities bordering seashores. Droughts can occur in any country like that in Brazil in 1983 or Mozambique, Africa in 1994. In California, USA, drought has caused miles of wild fires as wide as 65,000 square miles and two dozen fires simultaneously happening at the same time.

How pathetic that no immediate and instant solution has been made to remedy such gargantuan disasters. The Unified Field, when the threshold number is reached, the effect is instant and immediate. Like pouring cold water in boiling pot of water, the result is instant or as soon as the invincible shield is spread over, the result is instant. The Unified Field has a natural cooling mechanism and a character of an invincible shield that has been validated more than fifty times. I was not alone validating the Maharishi Effect. Many prominent scientists, most of them are Americans in Fairfield, Iowa and at the Maharishi University of Management who have conducted research and validations to proved Maharishi's theory of the Unified Field has an uncontestable results. I just duplicated the effect in small scale and found the effect instant and immediate. How awesome to remember the spectacular results from the few endeavors I have done in the past.

Remedies to solve climate change by governments have time tables, like 2030, 2050 and beyond. Meantime, climate change destructions will be occurring somewhere in the world. An instant and immediate solution has been found brought to light by Maharishi Mahesh Yogi for his discovery of the Unified Field of natural law. It is easy to form without mitigating companies manufacturing fuels, just by forming a group of TM and Yogic flying in large groups together will provide immediate and instant relief. The CO_2 that has concentrated in massive form in the atmosphere can be pushed away by the invincible Unified Field created and disperse into space as soon as the threshold number of TM and Yogic flying participants is formed.

The Global Country of World Peace has been established by Maharishi Mahesh Yogi before he passed away. It is a foundation that Maharishi formed in October 8, 2000 to provide an effective solution to mankind's perils. Maharishi Mahesh Yogi was the founder of the worldwide Transcendental Meditation movement and forerunner of natural law's principles of invincibility.

The Global Country of World Peace acts like a solutions agency of the world. Maharishi has brought to light the wisdom of the ancient Vedic science that contained the knowledge of Transcendental Meditation. The Maharishi Effect is an effective invincibility solution to the many problems of humankind.

Scientists working for the Global Country of World Peace interpreted the ancient Vedic science illustrated by the discovery of the Unified Field applying the most advanced scientific understanding of the basic functioning of nature in its seed form that can provide total invincibility from human suffering.

Chapter 7, the science of particles behind climate change and in Chapter 8, the discovery of the Unified Field was discussed in the simplest narrative in layman's format explaining the understanding of the invincible character of natural law which can be used to solve problems of man.

To summarize, this is how the Unified Field works. We have seen that the four fundamental forces of nature, electro-magnetism, weak force, strong force and gravity were discovered to unite at the level of the most silent, quantum state at the seed form of natural law and the state of unification scientist now call the Unified Field. Albert Einstein sought for it during his lifetime but was not able to locate it until he died.

Maharishi Mahesh Yogi was a physicists, he was able to confirm that the point of the unification of the four fundamental forces of nature is located in the human mind during the practice of Transcendental Meditation, the procedure of quieting the mind. TM was contained in the ancient Vedic text and now brought to light and interpreted by modern science. The Yogic flying illustrates the maximum brain coherence when the body lifts up defying gravity. It illustrates that the mind can handle gravity and unite with the other three forces of nature. Consciousness at the level of silence is the junction point of unification including gravity illustrated by the flying capacity of man hidden for many centuries, found and expressed in the ancient Vedic literatures. This flying ability of man locates itself at the main switchboard of natural law, at the Unified Field and can be handled by the human mind. This principle of flying naturally baffled many scientists and the general population, naturally it was met with skepticism. Flying was unknown for centuries, although in the sixteenth century saints in the Christian church started to levitate.

Maharishi has invited physicists to prove his discovery of the Unified Field. John Hagelin, a Summa Cum Laude graduate from Dartmouth College and a PhD holder from Massachusetts Institute of Technology and who has previously worked at CERN, the world's largest particle accelerator laboratory located in Switzerland joined Maharishi and corroborated that Maharishi's Transcendental Meditation and the Yogic flying technique can indeed create the point of unification at the level of the Unified Field when the human mind reaches its most silent state during the TM practice. Hagelin confirmed that the Unified Field is enlivened in pure consciousness at the most silent level of the human mind.

Eventually, many scientists and doctors joined in the events validating the Unified Field Effect now called the Maharishi Effect in honor of the late Maharishi Mahesh Yogi. The largest experiment was described in the Global Taste of Utopia event conducted in Fairfield in the late 1983 described in Chapter 6 and in many other events that followed truly validated the super-radiance, the invincible group effect described in the Unified Field discovery.

The electric field describes the behavior of electron and proton fields in physics. The Physics Chart in Chapter 8 describes the level of the atoms and particles that create the different fields much like the magnetic field, gravitational field and ultimately the Unified Field, found to be the home of the finest particles of nature.

Atoms break down to finer particles like electrons to protons, to photons, to quarks, to fermions and to finer particles which produce powerful natural force fields. The finer they break, the more powerful field they produce and when they act as a condensed particles they exponentially create a force field impenetrable and invincible from outside interference.

The field that they produce is a natural phenomenon of a state of invincibility and impenetrability. Just like the magnetic field produced by the magneto particles of the earth, it is impenetrable from the powerful solar flares from the sun. The magneto particles that surround planet earth from north to south provide an impenetrable invincible field - the magnetic field that shields the earth from the excessive heat from the sun. The graviton particles, that is finer than the magneto particles create an all powerful gravitational field that holds all the massive planets in place and does not fall from their orbits in space. Man does not fall in space because of the powerfully existing gravitational field around the earth. Gravity pervades the whole universe and they make all the planets move silently in place in an orderly, harmonious manner.

The particles at the level of the Unified Field becomes the finest particle and million and million times more powerful than all the other particles known that is why it has become the most invincible and impenetrable field and cannot be disturbed by any outside influences like climates, wind turbulence and other negative trends occurring on earth.

The particles at the Unified Field that I referred to as the AS_U particles, the anima solo particles that animate the whole universe appears to be the same as Maharishi's "sap" of the universe that creates "by itself, from itself and within itself" and create all the diversity in the universe.

It is the finest sole particle in the universe that acts alone not like the electron particle that needs to have a quark particle or the water particle that needs one hydrogen particle and two oxygen to produce the water particle.

Since the AS_U particle is the finest of all particles of nature, it is the purest seed form of natural law that neutralizes all negativities in the environment. Some scientists call it the God particle that produces only the positive aspect of nature. These cutting-edge scientific principles of invincibility from physics explain how in modern scientific terms enliven total Natural Law-the Unified Field- and its invincible qualities in the collective consciousness of society.

Maharishi Vedic Technologies of Consciousness create indomitable strength in national and global invincibility proportionate to the number of participants in the TM-Sidhi or Yogic flying programs that Maharishi has introduced during his lifetime.

The results of the super-radiance, the invincibility effect by the Unified Field effect was confirmed by extensive published research and a proto-type model is illustrated by the positive trends of life in Fairfield, Iowa, USA where the Maharishi Effect was experimented in a large scale in 1983. Fairfield has thrived nearly forty years with groups fly regularly daily in the Golden Domes located at the Maharishi University of Management.

All these fundamental principles of invincibility - the principle of indomitable strength, the principle of impenetrable coherence much like the Meissner Effect in magnets, and the principle of infinite adaptability, the gauge super-symmetry in physics become fully functional in the nation and the world when the qualities of the Unified Field are enlivened in national and global consciousness through large groups of coherence-creating experts practicing the Invincibility Technology.

Among the advanced techniques of Transcendental Meditation is "Yogic Flying," an ancient but profound mental technique that generates powerful, sustained brainwave coherence. Regular practice of Yogic Flying brings growing mastery over the force of gravity, as demonstrated by the body's spontaneous rising off the ground defying the force of gravity in response to a purely mental impulse.

Mastery over gravity demonstrates the capability of the fully developed human mind to function from the level of quantum gravity - the only level from which conventional, classical gravity can be controlled. The scale of quantum gravity is the Planck scale - the scale of Super Unification, where gravity unites with the other fundamental forces to become the Unified Field, see Physics Chart in Chapter 8 in the discovery of the Unified Field.

Here the ability to function at the level of quantum gravity implies the ability to function at the level of the Unified Field - the field of total Natural Law. Therefore, the growing mastery over gravity seen during Yogic flying truly indicates growing mastery over all the laws of nature. It is from this most fundamental and powerful level of the Unified Field - the field of total Natural Law- that the practice of Yogic flying enlivens indomitable coherence and invincibility in collective consciousness for nations and the world.

The Unified Field particles I previously referred to in Chapter 7 in the elementary particles of nature, behind climate change, the anima solo, the $AS_{U,}$ the particle that animates the whole universe is the most powerful particle far beyond photons, leptons, quarks, above and beyond the power of the atoms. And because it is the finest particle of them all, it creates the greatest impenetrable field that shields the earth against negative trends. It is a phenomenon in nature - the finer they become, the more powerful field they produce!

The AS_U particle can produce the greatest power on earth as it is in the power of the heavens and the universe, when concentrated in form, it creates positive trends on earth. The Unified Field is unlike the power of the atomic bomb made of concentrated hydrogen particles in nuclear that wiped out a whole city in Hiroshima in Japan that ended World War II. Concentrated hydrogen particles contained in tons blasted off destruction or it is like the monster group of concentrated wind particles, so powerful that devastated miles and miles of communities at the central island of the Philippines displayed by the Haiyan typhoon.

The wind particles in the super typhoon Haiyan swallowed several towns along its merciless path leaving widespread death and devastation. The Unified Field does not destroy or create havoc, it nourishes providing an impenetrable shield from negative influences in the environment.

Like the magnetic field shielding the earth from excessive heat from the sun, the Unified field produce a much stronger exponential field that shields the earth from powerful wind turbulences, that was why the 1,200 meditating flying participants that Maharishi had sent to the Philippines in 1984 and the 500 students practicing TM together that I had repeatedly formed that followed at that time in the Philippines resulted in climate turbulence fading away and removed the negativities that then prevailed in the country.

It was an amazing phenomenon, the four hundred meditating workers that built the thirty-story tower (see photo in following pages) created an impenetrable field that prevented the fires that broke twice on both sides of the tower from different adjacent properties, the fires did not create a single dent from the fires that had occurred.

The Unified Field created a shield formed by the four hundred meditating workers prevented any single timber in the tower to burn. Not a single flame burned any part of the tower. The invincibility and impenetrability created by the four hundred meditating workers shoved off of the fires that gutted the other adjacent structures to ashes piled on the ground and not a single timber in our tower burned to everyone's amazement.

The Unified Field is indeed impenetrable, not a single piece of wood burned at the time we were building the tower. It was indeed awesome with 400 workers witnessing the powerful invincible field produced by the four hundred meditating workers enlivening the invincible Unified Field lingering within the human mind.

Now a super-radiance group needs to be in place in the Philippines and in every country equal to the square root of one percent of its population practicing TM and the TM-Sidhi together in one place to create natural law principle of invincibility from any negativity occurring.

Typhoons that frequently blast and the greatest super typhoon ever recorded in the history of the Philippines that made a touchdown would have been prevented by a super-radiance effect of the Unified Field at the level of the human mind. It will create an impenetrable field, an invincible powerful Unified Field that will shield and protect nations from fires and storms. Storms and hurricanes will just instantly stop with the super-radiance group formed..

The fifty years of the life of the late Maharishi Mahesh Yogi proclaimed the nourishing aspect of natural law to the whole world and Global Country of World Peace, www.globalcountry.org continue his precious heritage until people listen instead of wait for another record wrath of nature blasting off another place somewhere. Who knows which place and who knows who will be its next victim.

I remember before the fall of then President Marcos in the Philippines, we had an audience with a group of generals in the military called for by then President Marcos. I spoke about the nourishing effect of the Unified Field and its application at the level of silence within the human mind.

One general raised his hand and said, "I learned Transcendental Meditation. What they are saying is true." The rest of the generals in the audience, they were the elite in the military, they just stared in disbelief. They could not comprehend the science of the Unified Field and ignored us. One commanding general even laughed at me explaining the significance of Transcendental Meditation.

You just cannot laugh at nearly three hundred wide miles of horrifying devastation and death brought about by the Haiyan super typhoon that crossed both ends of the country. No one has any right to ignore such vast devastations without considering finding a solution to these magnitudes of destructions however unconventional they appear to be, and because I found it to be true I want people to know that it has a scientific basis.

The Unified Field is so powerful, more than enough to shield the earth from destructive climate disturbances. It can be drawn at the silent level of the human mind in large groups together during the TM and TM-Sidhi programs that Maharishi has brought to light during his lifetime.

It is best to review and analyze the scientific basis rather than expect another monster storm – Haiyan, Sandy, Katrina, the tsunami in Japan or whatever name it will be called to happen in the future. These climate havocs need not happen anymore as the technology is now available and is at hand.

The Unified Field has been validated by several documented experiments from distinguished scientists and confirmed it to be effective in neutralizing negative trends occurring in the environment.

As a civil-structural engineer, I had some training in science and classical physics. I was just a single mother wanting to protect the future for her children and who experimented the Maharishi Effect for my personal reference, not only ones, but repeatedly. I have seen the evidences first hand. I know it is true. I agree with Maharishi, heaven can now be created on earth. The discovery of the Unified Field defies all sciences because it is the supreme science that can transform and create a better life on earth. Instead of just witnessing wild typhoons, flash floods, freezing climates and wild fires regularly occurring happening here and there somewhere around the world, listen to Maharishi's urgent call, "Do it. Just do it." Heaven is possible on earth.

Balancing climate was beyond modern technology to understand because particle science is a new found science and the discovery of the Unified Field spanned only fifty years of Maharishi's life. He was the one who brought the Unified Field to light and still needs to be accepted by governments, though TM has gone mainstream in the USA, only two governments tried it, Mozambique in Africa and Cambodia in 1994. It took nearly one thousand six hundred years before Galileo's theory of sun centered universe was accepted. Maharishi's Unified Field has been validated for nearly fifty years and many scientists have confirmed it to be true.

Thermodynamics express the behavior of air particles by changes in temperature. The rotation of the earth around the sun determines the seasons on our planet earth. In summer the earth is nearest the sun so the season is warm and hot or in winter the earth tilts farthest from the sun so the season is cold.

Scientists stipulate that the ozone layer covering the earth from the sun is being depleted because of the carbon dioxide particles being emitted in the air from man-made carbon waste, thereby increasing the earth's temperature causing the melting of the ice in the north pole, increasing ocean temperatures and producing climate change. Abnormal lengths and cooler freezing temperatures is happening on earth , it is caused by the cooling of large concentration of greenhouse gases in the atmosphere as the earth tilts away from the sun.

It was stipulated that the depletion of the ozone layer increases the earth's temperature causes the imbalance in the atmosphere. The magnetic field of the earth shields the earth from excess sun's rays, so how can the ozone layer cause increase in the earth's temperature? The reason is the CO_2 layer when heated, it evaporates the air in the ozone layer. This theory about climate change, the cooling and invincibility character of the Unified Field recently discovered quantum in physics can solve global warming.

The Global Country of World Peace have leaders who are experts in the technology of the Unified Field in nearly all parts of the world. Experts can be immediately called upon to support any country any time, in any place in the world.

The most scientific concept of balancing climate changes is held by the Unified Field which is invincible from outside negative influences. The Maharishi Effect is an instant solution to climate changes just like Maharishi has said!

Heat or cold conditions affected air turbulence. When there is more heat in the atmosphere the more movement of cold air becoming hot or vice versa, the more cold air in the atmosphere the more hot air becoming cold. When balance is not maintained like when there is more cold air, it condenses to more clouds producing more rain and results to floods. The reverse happen when there is more heat in the air, the cross collision of hot and cold air particles create thunder and lightning sometimes producing fires.

Since the time of Albert Einstein who stipulated that the three original concept of the atoms, the protons, neutrons and electrons were not the fundamental particles of nature. Einstein discovered the formula to contain the hydrogen particle and eventually the hydrogen bomb was developed. He described that quantum gravity displays the character and properties of the graviton particles contained in his theory of relativity

Furthering the breakdown of atoms to finer particles, hundreds of particles followed, eventually the finest form of particles was found to exist at the Unified Field occurring at the silent level of the human mind during the practice of Transcendental Meditation.

At the level of the Unified Field, the $AS_{U,}$ the particle that is indivisible and contains the seed form of nature will interconnect with all the other particles in the earth's atmosphere creating order and harmony in nature. When there is turbulence in the human mind it follows that turbulence in the environment also follows because every particle in the universe interconnects with each other. Like the light particles, the electrons, moving around to light the whole house or photon particles from the sun lighting our whole universe. The AS_U, the anima solo particle that animates the whole of the universe present in the human mind also permeates the whole galaxy. All the particles pervading in the whole universe are all interconnected together including the AS_U particle lingering in the human mind.

Like a satellite message sent from the International Space Station thousands of miles away from the earth reaches the planet earth in an instant. So when the human mind attains that quietness at the level of the Unified Field the whole surrounding field attains that quiet balance in the air maintaining climate temperatures.

It is a scientific truth that the magnetic field that envelopes the earth shields the planet earth from excessive heat of the sun particularly from solar storms that blasts the earth and like the Meissner effect in physics, the magnetic field throws off negativities that come in contact with the magnetic field.

This theory of the earth's magnetic field shields the problems of solar storms and protects the earth, it has been confirmed by scientists since its discovery during the past several decades.

Storms and winds particles within and below the clouds and fire particles that constantly bombard the earth have an invincible shield for their attacks much more invincible than the magnetic field created by the magneto particles and because the AS_U particles are million, million and million times finer than the magneto particles, the Unified Field that they create becomes truly invincible. Therefore, it makes the field produced by the AS_U particles become the most invincible of all fields, on and above the earth - the Unified Field. Any negativities or any turbulence above and below the earth's atmosphere cannot be penetrate and disturbed by this invincible Unified Field.

The invincibility and strength of the Unified Field is proportional to the number of human brains containing the finest AS_U particles put together at the level of coherence during the practice of TM and Yogic flying , it determines the degree of invincibility. As of 2015, 9,000 is needed for the whole planet earth and 2,500 for the USA. The bigger the population, the bigger number required.

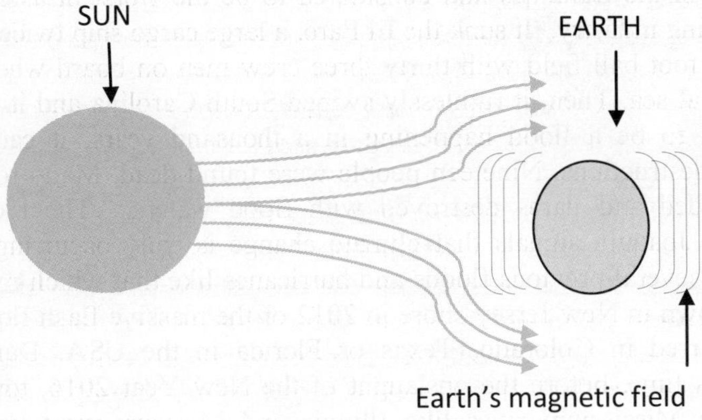

MAGNETIC FIELD OF THE EARTH

Maharishi's Tower of Invincibility

I concluded that the Unified Field effect in action shielded our building from two adjacent fires. Not only did the Unified Field shielded our building from fires but the several rains that occurred stopped the storms and prevented the floods around the area.

Our building was just a block away from a regularly flooding section of the city that floods each time a heavy downpour occurs. Each time the rain would pour down, I would asked the workers to stop work and meditate. The rain stopped immediately soon as the workers opened their eyes after each meditation. Every time the workers meditated, the rained stopped. It was just too awesome to watch the repeated coincidences happening. It was just too obvious to the astonishment of the 400 participants! I concluded that the Unified Field effect worked each time the workers did their meditation and in groups together thinking the same desire, the intent is fulfilled.

Thru the group effect, the Maharishi Effect, climates balance and the whole environment is nourished with a perfect condition which prevents typhoons, floods and fires from occurring.

And because the Unified field is the field of harmony and order in nature, it follows that if a large group practicing the TM and TM-Sidhi together in one place, harmony and order on earth will be created. Heaven on earth will prevail like what Maharishi has foretold.

It is therefore the responsibility of every government to test and validate the Maharishi Effect and create a permanent group of participants of the TM-Sidhi program that Maharishi has brought to light.

As the magnetic field that surrounds the earth shields it from sun's excess heat, the Unified Field shields the earth from any negative trends occurring when a group of 9,000 TM-Sidhi participants are placed together in any one place on earth, then positive trends exponentially increase. All non-good vanish and only good trends prevail.

Science describes that finer particles create exponentially much greater strength. The Unified Field particles are much finer than the magneto particles, hence they project exponentially much greater field of invincibility.

With so many lives lost and billions of homes and infrastructures destroyed by floods and fires, this knowledge cannot anymore be ignored no matter what skepticism exists in the general population.

Just imagine the horrifying scenes in the Philippines happening after the Haiyan super typhoon. Would you rather approach a TM center or call the Global Country of World Peace? Would you rather call your government leaders and tell them of the solution?

Like the New Jersey shore that was rampaged by a devastating storm, then burned down again. Fifty buildings were just re-built after the Sandy storm then mercilessly burned down again. How tragic indeed!

In 1994 in Mosambique, Africa at the time of then President Chisano, in an area where there was drought for twenty years, rain fell, when large groups in the military started to meditate. The climate balanced. In Brazil, during the Global Taste of Utopia event in 1984, rain fell in drought areas.

In the two schools I had established, it rained soon as the students started to meditate twice, in the two separate schools, it cooled and balanced the warm climate. In the thirty storey tower, fires broke out and not a single lumber burned, but the two adjacent properties burned to ashes piled on the ground. Storms broke out but stopped soon as the workers started meditated. All the 400 TM participants were just stunned with awe by the Unified Field effect.

These scientific analysis and the above events really made my belief in the Maharishi Effect confirmed to be true after the large groups that I had formed with meditating participants created the impenetrable shield that protected the country from storms and from other negativities occurring at that time of history of the country.

Maharishi seemed to be the only one who has proposed a solution to balance climates when no one has ever come up with one instant solution. Unconventional as it may seem, it worked all the time! Before he closed his eyes and before he died, Maharishi's last words were, "May the world live long in fulfillment, peace, happiness and prosperity."

The Unified Field effect, the Maharishi Effect can be in place anywhere in the world. In Fairfield, Iowa USA, Maharishi heritage, the seed of his knowledge was planted there for nearly forty years. His leaders at the Global Country of World Peace are custodians of his precious heritage and they can provide the knowledge to create peace, fulfillment and prosperity.

15

MAN IS THE CONDUIT OF NATURAL LAW

During Maharishi's lifetime, he expanded the state of temporary transcendental consciousness, experienced during the twenty-minute practice of Transcendental Meditation leading to a permanent state of cosmic consciousness.

When the mind reaches the state of the most subtle quietness, a restful alertness begins to be experienced at all times-even at the waking, dreaming or sleeping state, at the same time the nervous system attains the most refined and balanced functioning after some time of constant and consistent practice of Transcendental Meditation and Yogic Flying, the constant feeling of bliss becomes a constant state too, then, cosmic consciousness becomes a permanent state. Eventually, all the seven states of consciousness from awake, sleep and dream state, progress to cosmic consciousness, God consciousness and unity consciousness and to which describes permanent settled states of quiet awareness occurring at all times, even awake, sleeping or dreaming..

The body, however, needs to be rested at all times. Any activity that tires the body reduces the state of bliss. Mind and body are interconnected. Tired body makes the mind dull and imbalances behavior. TM provides deep rest twice deeper than sleep while the mind is alert.

Maharishi once advised me when I complained that I had so many things to do. He told me, "Rest, sleep more, and let others do the work for you." Now I understood more what he meant about seeing to it that the body is fully rested at all times.

Guru Dev, Maharishi's master has said the same. The body needs to be cared for well enough because it takes four million processes of nature to perfect it. The body needs rest to be a conductor of the Unified Field simultaneously occurring at the silent level of the human mind. Eventually, when cosmic consciousness is attained, that means transcendental consciousness has progressed from temporary to a permanent state, the mind attains full enlightenment, the body gains perfect health and the feeling of bliss stays constant and overlaps the other states of sleeping, dreaming and awake states.

Many scriptures state that the body is the temple of God. Cosmic consciousness overlaps all the other states and surpasses the temporary state of silence, the transcendental consciousness experienced during the twenty minutes of the Transcendental Meditation practice.

The temporary state of transcendental consciousness during Transcendental Meditation gradually produces a refined transition of the human nervous system to the level of cosmic consciousness sharpening the level of the mind and culturing the functioning of the nervous system to a permanent state of perfection, to cosmic consciousness. When both mind and body are in the state of cosmic consciousness, life begins to be lived in fulfillment that means, two hundred percent, one hundred percent inner happiness and one hundred percent outer affluence. It is like a tree that has rooted down deeply and reached its peak of growth and starts to bear fruit. Cosmic consciousness is the peak of perfection in man. All thoughts and actions become in accord with the evolutionary direction of natural law. Like a tree, constantly watered, the tree grows and bears fruit. Fulfillment in life reaches the level of perfection in cosmic consciousness.

In man, with the constant and consistent practice of Transcendental Meditation, drawing energy from the Unified Field, a life in fulfillment is spontaneously reached. Maharishi calls this state of fulfillment, cosmic consciousness. In time, it becomes a permanent state.

Maharishi describes this state in this analogy, "It is like a white cloth constantly dipped in a yellow dye and brought to the sun to dry. With constant dipping and bringing it to the sun, the white cloth permanently absorbs the yellow color and the yellow dye does not fade away anymore."

The same thing happens with the mind and nervous system, with constant and continued process of dipping into that root energy field, the mind and nervous system permanently absorb the Unified Field, then the nervous system becomes a permanent channel of the Unified Field. Man becomes the conduit of natural law. His thoughts, words and actions follow the path of natural law and become an instrument of natural law.

Like the copper wire, which is a natural conductor of electricity, man at the level of the Unified Field becomes the conductor of the cosmic energy, the Unified Field, the root energy of the universe, then life becomes in accord with nature.

To become an instrument of natural law is the goal of all religion. When all pursuits become the pursuits of happiness for all men and for the good cause for humanity, God consciousness becomes a permanent state too.

God realization is a state and condition of expanded intelligence and refinement of the nervous system simultaneously occurring. Both mind and body become free from stress and both function perfectly at the most balanced state. This is the purpose of religions, God realization. In contrast, many religions express God consciousness by their interpretation of books, like different interpretations of the Bible or the Koran. "Our book says this (or that), we need to follow," these are common assertions on God and thus govern their laws man is made to follow even when stress that has accumulated in the human nervous system and makes man unable interpret what is right or wrong and blindly follow the laws religions interpret.

In contrast, Maharishi describes God consciousness as the refinement of the nervous system and the expansion of the human mind. No wonder many wars are caused by religion! Instead of expanding of the mind understanding truths of natural law, many still follow narrow-minded postulates. Instead of refining the nervous system, eliminating anger, greed and the ego, stress continually accelerates causing constant confrontations and conflicts. This is what happens in turbulent nations like in the Middle East, without a popular and acceptable stress reducing technique like TM, diverse religions and conflicts dominate!

Maharishi's Transcendental Meditation is not a religion but a technique to effectively expand the mind and removes stress that is the main cause anger and greed. When God consciousness is finally reached thru Transcendental Meditation and Yogic flying, enlightenment is obtained and the ego fades away. Man automatically concerns himself with service for the good of humanity not any more for his self interest alone. Service to benefit other men besides himself becomes his prime purpose. Man then draws the qualities of the Unified Field and becomes the conduit of natural law. In large groups together, man has the capability to balance the climates, create peace and prosperity. Man becomes master of natural law! Yogic flying accelerates time to reach God consciousness, it is faster path than the basic TM technique and one hundred times more powerful.

This is Maharishi's interpretation of the scientific approach to God. It is the state that another physicist, Lecomte referred to, as the perfectly evolve man. In the book by Maharishi, "The Science of Being and Art of Living," published in 1963, he said, "It is a very delicate path to realization." He was referring to God realization, God consciousness. Maharishi discussed five approaches to God realization describing the process in the following manner: intellectual, emotional, physiological, mechanical, and psychological approaches to realization. The different process suits the individual's preference, experience and degree of stress.

Like a tree, the leaves realize the unchanging nature of the sap within the tree or the fruit realize the permanent unchanging nature of the same sap within the same tree. Like different personalities with different attributes of mind, heart, and nervous system, they all have the capability to realize the permanent unchanging nature of God in the different manner that Maharishi described. His Transcendental Meditation makes it easier and simpler for anyone desiring to reach God consciousness and be a channel of godliness. And because man has a body, a mind, and a nervous system, all men are capable of reaching God consciousness. God consciousness is reached thru perfecting the functioning of the mind and nervous system without stress remaining.

In the scientific approach, it is like a glass, a window, when all the dust is removed, you see clearly what is outside. When the nervous system is free of stress, you realize God! You experience the blessings of God in your life. Like the copper wire, the electric current passes through the wire because the copper wire is the natural conductor of electricity.

So too with man, with expanded mind, healthy body, and refined nervous system, he becomes the natural conductor of the finest energy field, a natural channel of the Unified Field at the level of silence at all times and in all states, sleeping, dreaming and awake in the state of permanent God consciousness.

In the terminology of religion, you may call it the field of divinity, or the field of sanctification, or call it whatever name, so long as it brings bliss within, expands the mind, makes the body healthy and improves the environment, it becomes what it is, an unseen energy source.

It was written in the Christian Old Testament. "I Am who I Am," an invisible God. Someone asked me one time, "Is it the Holy Spirit?" "Call it whatever name, so long as you can see that it improves the mind, body, behavior, and produces positivity in all aspects of life, balancing climates, creating peace and prosperity." I repeated the same answer. In God consciousness, the spirit within you becomes holy. You have drawn the attributes of the Unified Field of natural law. You can get support of nature constantly.

"Like a tree, with water, it grows and bears fruit to serve its purpose. You will know the tree by its fruit. The important thing is to experience it. It is a must-do process. Draw energy from it. Twenty minutes in the morning and twenty minutes TM in the afternoon. The process is taught by the teachers of Transcendental Meditation who were trained by Maharishi during the past fifty years of his life. There are nearly forty thousand TM teachers scattered around the world.

It is not difficult to practice TM. It is easy and effortless, even a six - year-old child can do it. If you can harness the total potential of natural law, wouldn't you do it? This concept of being able to draw the total potential of natural law is easy. It is like when you take a bath, you do not need long hours to clean your body in the shower. After you have cleaned your body, you go out to attend to your daily activities. It is not necessary to stay longer than twenty minutes in Transcendental Meditation. Scientific results showed that twenty minutes of TM practice is sufficient enough time to get the energy that last thru out the day and to be able to perform all the necessary actions for success.

Becoming the conduit of natural is to posses total potential of natural law that is within reach by the human mind. And in large groups together becomes master of natural law capable of transforming man's environment that can balance climates, create peace and prosperity to prevail. It should not be misunderstood, however, to meditate long hours because you want to realize God. No, it is not the long meditation that tires the mind, like in the ancient and medieval monasteries that have now become empty because of long, tiresome and seldom effective meditations. Transcendental Meditation differs from other forms of meditation in ways that it immediately makes you feel relaxed and happy in an effortless manner, in contrast to different concentration techniques that may take longer time and tires the mind.

Twenty minutes of the TM practice is enough drawing energy in the morning so that you have energy to perform successful actions needed during the whole day. And in the afternoon when the nervous system and body become tired after work, you go back again twenty minutes, relax, and release all the stresses you have accumulated during the day.

The actions performed after meditation, are actions that are needed to be performed during the day. They become actions done with more energy and with clearer vision. It is not just meditating twenty minutes and waiting for things to happen. Right action after meditation is important for success. Not any more being a recluse in a monastery, but happily active in the outside world.

Monasteries have become empty because of the long old religious meditations that were tiring. Besides living the ascetic and celibate life is difficult and appear impossible to maintain. There is so much effort to maintain compliance in a monastic life in contrast to Maharishi's concept of two hundred percent of life. He said, "Live one hundred percent inner happiness and one hundred percent outer affluence." TM is easy to practice and is blissful.

Scientific validations say that twenty minutes is sufficient enough time. That is all the energy you need to work and to go into your activities of the day. Life should be lived with both, inner peace and outer happiness. At the same time, you live your life in affluence, not asceticism like most believe. Maharishi repeated this every time, "Inner peace, outer affluence."

In time, the condition of the mind and nervous system progresses to a balanced and more refined functioning, the twenty-minute transcendental state becomes a permanent state of cosmic consciousness. Like a tree, not only the trunk has grown tall and big, or the leaves turned green, it has to bear fruit to complete its full life cycle and purpose. Man has to reach this purpose of two hundred percent of life to also complete and perfect cycle of man's life.

Maharishi, in the beginning, after he finished his physics and mathematics degree, he tried to bring the pieces of God together, like a puzzle he could not complete. There were some pieces missing about the concept of God. Maharishi's quest for the answer why suffering exists and why if God is all good, why does he allows suffering occur? Like all the brilliant scholars who feel some kind of void, some emptiness and yearning for more answers in their quest for God. Why is there suffering? Why so many system of belief? In the Vedic literatures, happiness was extolled, not suffering.

I had similar quest about how the Unified Field can remove the common trends of suffering and one that holds the secret to happiness? What is the Unified Field? Where do all the good things from the TM practice come from? Where do these positive happenings change the common scene of sufferings we constantly see occurring around?

The science of particles opened it all up to me. People begin to question, "Who am I? What is the purpose of life? Why are there so many people suffering?" These are the same curiosities people ponder, especially in a time of need, when problems and suffering present themselves in one's life.

When there is no one else who can provide support, most people seek God and religion and hope that they choose the right one, otherwise suffering will still exists, like there are religions that cause terror and conflicts, Now, "There is an unseen energy field we can draw from, the seed energy, the Unified Field. God created it for our use. If God is all - merciful and good as all the religious precepts say, how come all these calamities occur? How come these negativities like wars, human cruelties, hunger, and want among the people persist continuously? There must be something missing, some knowledge hidden somewhere about God.

This was the reason Maharishi embarked on his journey to seek out the missing knowledge. Maharishi decided to go to the deep forest of the Himalayas and seek out a teacher who could open the doors of enlightenment and know more about the missing attributes of God. Maharishi travelled in the rugged mountains of the Himalayas searching for someone who could teach him about the missing puzzle of God. As soon as he saw Guru Dev, his search for a master ended. The moment he saw, Guru Dev, his teacher, Maharishi said, "One glimpse of him, I knew he was the one." He surrendered his life to Guru Dev. Maharishi served Guru Dev for more than twelve years, right to the very end of Guru Dev's life. Guru Dev passed away in 1955. Maharishi described his life with the master he served, Guru Dev, saying, "I followed his every move to see how I can serve him best. I tried to anticipate his thoughts and followed the direction of his thinking. I did what I thought he was thinking."

"He was thinking this and thinking that. I did this and all that I thought he was thinking. Soon, everything became so easy, so effortless." The science of a group or a pair participated with Maharishi's thoughts and the thoughts of Guru Dev, his teacher. Two minds working together resulted in easier action and greater fulfillment with exponential results. The mind is the seat of the unseen, unmanifest energy of nature now brought to light as the Unified Field of modern science.

Just like in a good marriage, either of the couple should do the same as Maharishi did with his Master Guru Dev. Couples thinking together, moving forward exponentially together, attain their goals more easily than when thinking and moving alone. Like a car that has stalled. Pushed by one person, the car will not move forward. Push the car with more people, it will easily move.

By Einstein's formula, $E= MC^2$ force equals mass times the square of acceleration. Thought force and energy operate in a similar exponential power. Thought force is strongest at the level of the

Unified Field. In the physics chart at the Planck scale illustrated in Chapter 8, in my interpretation of force field energy, showed the exponent appearing to be to the 33rd power. The Unified Field chart was a puzzle to me until Maharishi's theory was validated in 1984, when Maharishi called for TM and Yogic flying practitioners from around the world to go to the Philippines, at that time the country was struggling from many negativities occurring. "Go, as if the house is burning," Maharishi said.

It was a time in the country, when a volcano was erupting, a strong, wild typhoon and an earthquake simultaneously occurring in the same year. There was this turbulence amassing in great numbers among the people yelling in the streets of Manila! Multitudes of people were angry. Millions were rallying in the streets of Manila against then President Marcos. Maharishi knew that the level of the mind is an energy field more powerful in groups together and necessarily needed a square root of one percent of the population to counter the turbulent minds. Like Einstein's formula $E=MC^2$ where two is in the exponential square position. Maharishi sent 1,200 Transcendental Meditation and Yogic practitioners to quiet the turbulence. As soon as they did TM in large groups together and quieted their minds, the turbulence subsided. Like a boiling pot of water cooled with the equivalent volume of cold water in order to instantly counterbalance the heat. It was all a mathematics theory exponentially increased in number!

It was like the multitude of collective angry minds, which became so powerful in energy, seemingly caused so great a force as to shake the earth with an earthquake of great magnitude at the same time causing fire and brimstone in the volcano that was erupting or the wild wind blowing in the storm simultaneously occurring. The one thousand two hundred TM and Yogic flying participants who came to the country, calmed the turbulence, the volcano and the earthquake simmered down, the storms faded away and there was a peaceful transition to a new government without firing a single shot. The energy was more than the energy beyond the atoms at the level of the Unified Field also lingering within the human mind in large groups together illustrated in the Physics Chart on Chapter 8.

It took some time for me to understand Maharishi's thinking. But eventually, Maharishi's Unified Field and Physics Chart started to make sense to me. I began to observe world events, people rallying in China during the students protests in Tianamen Square, an earthquake shook China. When monks were killed in Tibet in

Myammar, people protests rose in the streets, an earthquake and typhoon happened.

When a volcano erupted in Iceland, I was curious. I read that there was an exodus of Icelanders. People were angry. Their minds were in turmoil. It was the first exodus of Icelanders in a century because they were angry with their government. Maharishi did not leave assistance to chance when the same occurrences happened in the Philippines in 1984. He sent the remedy.

I began to understand Maharishi Maharishi's Unified Field principle. We are now in the scientific age, and capable of understanding occurrences in nature. And because the human mind contains the seed form of nature and becomes the conduit of natural law, the answer can be located within the human mind.

After Maharishi passed away, I confirmed that he was the greatest scientist who found the missing piece of the puzzle between man, nature, and God. God created man in all perfection. The knowledge of perfection was lost and is again restored for man to draw from through TM. God's missing knowledge unfolds at the level of the seed form of natural law within in the human mind at the level of silence.

I give gratitude to Maharishi for this great knowledge of quieting the mind thru his Transcendental Meditation technique. I give gratitude to the unseen God who created the energies of nature. God has provided water for the plant to grow and to bear fruit and for man to drink.

God created the magnetic field around the earth to protect our planet from the heat of the sun. God created gravity for us not to fall off the planet earth. God provided us with the electrical field to light the earth and benefit man.

With the discovery of the root energy field, the Unified Field, from Maharishi's Transcendental Meditation, God has provided man an energy field to draw from, to grow, to be fulfilled and to live life without suffering. Transcendental Meditation is a simple mental procedure to quiet the mind and remove problems in one's life. God has provided man the capability to finally control climate change and balance the weather. God provided man a built in mechanism found within the quiet level of the human mind in large groups together to prevent typhoons, tornadoes and earthquakes. No gratitude is enough for this greatest of all wisdom.

God has provided a formula for wars and terrors to end, but was lost in obscurity for many centuries, Maharishi brought it back. No greater gratitude is sufficient enough for this great knowledge.

God has provided man a formula to blossom in health and wealth, wisdom and peace. God made man to the level of perfection and provided a tool to reach it thru the Transcendental Meditation technique.

I give gratitude to all the scientists who have brought us the knowledge of electricity, solar energy, quantum energy, and all the energies that benefit man that have been discovered to date. I wish for deeper knowledge, understanding and wisdom for the study of particle science and its benefits to man.

I bow down and give gratitude to Maharishi, my teacher, who has brought the knowledge of Transcendental Meditation that provides man the procedure to reach perfection and eliminate suffering. Maharishi made commentary on the ancient book of the Bhagavad-Gita. He compiled the old Vedic books and interpreted them. He refined Patangali's ancient yoga sutras and made it part of the TM program.

Maharishi has found that which was missing in God Consciousness or our consciousness about God. And because God is complete and perfect by himself, it is difficult for an imperfect man to truly understand a perfect God, unless the imperfect man himself rises to the level of perfection to fully understand God who is perfect. Maharishi's Transcendental Meditation brings man to the level of perfection, to the state of God consciousness.

TM becomes a tool for man to merge with God. Man is blessed by Transcendental Meditation, because the Transcendental Meditation technique is so easy and effortless that makes the path to God in the same easy manner.

The path to God is easy Maharishi style! Guru Dev, Maharishi's master taught him the ancient wisdom of the Himalayas. Maharishi brought scientists to validate the Himalayan knowledge empowering man. I read an ancient Vedic verse, older than five thousand years. It was written by some Vedic masters, in an era and period of the ancient Vedic civilization. I repeat the phrases I previously wrote. To me it opens a hidden powerful knowledge which all men should know. I repeat the verse;

"Go together, think together, know your minds to be functioning from a common source, in the same manner as the impulses of the universe remain created, united at the source." This ancient verse contained in the Rig-Veda confirms the group effect.

Maharishi must have been guided by this Vedic verse. He called his science, the technology of the Unified Field, "united at the source." I call it the root energy field, the finest solo particle, the anima solo particles in the universe. It is what it meant to me.

There are great values hidden behind the ancient books of wisdom, like the story of David in the Old Testament. David was a small shepherd boy, one among twelve sons, most of his brothers were great generals. No one of them became king. David, the small shepherd boy became a great king. A stone from a sling thrown from a humble shepherd boy blinded a giant. David defeated the giant Goliath. It was a battle of the small with the big!

Greatness does not come from a name that is great but from one who performs a humble service, like tending the heard of sheep for his father. David surpassed his brothers who were great generals. He became a king! "He who is the greatest, let him be your servant," Jesus Christ once was quoted in the Christian bible. Service is a requirement of nature illustrated in the Ego Chart attached. God consciousness is displayed when ones concern is not any more toward the self but towards service to other men, by then the Ego has faded away. Natural law takes over and his life leads to God consciousness.

I remember while I was in India enjoying the super radiance effect brought about by the 7,000 meditating pundit boys that Maharishi was organizing at that time, in 1988, I came across a book in one of Maharishi's waiting rooms. Inside the book, a chart caught my attention. It particularly interested me because it is a common character so often displayed, the Ego. I have observed that generally, credit to oneself comes first before any other.

The position of the Ego above the Unified Field fascinated me. See the Ego Chart in the following page. In FIG. 4 in the following diagram, it appears that the Ego blocks the Unified Field, the seed form of natural law. The mind and the intellect apparently spring from the Unified Field. Without the Ego the mind and intellect directly connects to the Unified Field.

The Ego blocks the direct interaction of the intellect to the seed form of natural law. The ideal position would be in direct contact with the total potential of natural law without the Ego blocking it and therefore support of nature becomes available every time. The Unified Field provides support of nature.

In my simple interpretation, to obtain a direct access to support of nature, the Ego has to be completely eliminated. The Ego blocks the Unified Field. FIG. 4 illustrates how the Unified Field, the seed of natural law interconnect with the five senses of man, the mind and intellect. The Ego follows after the intellect in the chart.

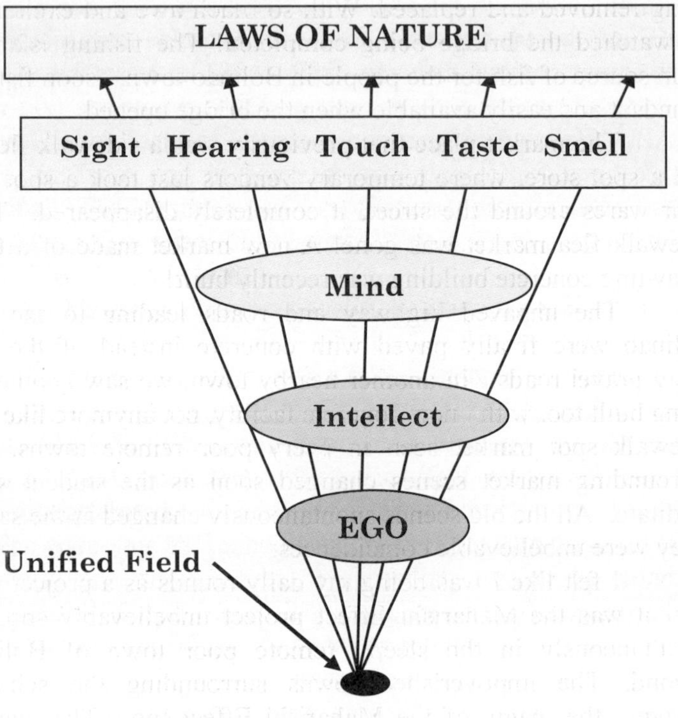

FIG. 4 Maharishi's Ego Chart

According to one of Einstein's biographies, "Beyond Einstein," he was expounding to go beyond nature and find out the purpose of the five senses of man in relation to nature. Einstein wanted to correlate and connect the five senses, sight, hearing, touch, taste and smell with nature. It appears to illustrate its connection with the laws of nature at the topmost position and the Unified Field, seat of nature, lies beyond the Ego. Without the EGO, the flow of the energy field goes above to the intellect, mind and the five senses, which illustrates that the five senses are the conduits of the of nature. The relationship of the five senses in Fig. 4 to the Unified Field and the laws of nature are likewise interconnected.

If the Unified Field is enlivened within the human mind, the five senses connect with natural law. That is why Maharishi said we are the master of natural law at the level of silence during the practice of Transcendental Meditation in large groups together.

The resistance of the energy field in electricity relates to friction, in the human nervous system acting as the conductor of the Unified Field, similarly, the Ego posses resistance. Friction within the electric wire reduces the flow of electrical energy. Friction is the resistance that blocks electric current from flowing thru the wire. Like electricity, once the bulb reaches its full capacity, any extra current that comes into the bulb, causes the light bulb to flicker out as it overloads the capacity of the bulb. The extra current has to move to another bulb, then to another bulb and so on to carry whatever extra current is flowing in the copper wire.

When the human nervous system is overloaded with stress, it blocks the flow of the Unified Field benefits. Ultimately, if the Ego has faded away, fulfillment of desire becomes spontaneous, there becomes automatic support of nature. Ego blocks fulfillment in life.

And because, in my interpretation, the Unified Field is the root of all the laws of nature and Maharishi said that if you serve the cause of natural law, natural law will support you. It is not anymore that you support yourself - natural law supports you. It is vice-versa, not yourself but the cause of others that will obtain maximum support of nature.

It is like Mother Theresa, supporting the poor. Many people supported her cause of service to the poor. Service to others draws spontaneous support of nature. In man, it looks as illustrated - Ego blocks the frictionless flow of the Unified Field, the total potential of natural law

In a copper wire, friction poses resistance to the flow of electrical current. Friction is a force that acts against the flow of the current. I related the resistance to the support of nature with the ego that acts against fulfillment in man acting as the conductor of the Unified Field. The less self-directed man is or egoless, like if you serve the cause of others, you will receive more spontaneous support of nature. Without the ego, man automatically taps of the root energy field and its attributes.

Beyond the self, lies greater power, the power of natural law. EGO acts against the flow of the energy from the Unified Field and the spontaneous support of nature that Maharishi was referring about.

The goal of man appears to be towards service to others, beyond the self, beyond the EGO. This appears to be scientifically true, as in an electric bulb, soon as the number of watts is reached in the corresponding size of the bulb, the extra energy has to move to another bulb, otherwise the bulb will overload and flicker out. So, as energy moves in the copper wire, it has to move from one light bulb to another light bulb, like that, further increasing the total brightness.

So too in man, as soon as one has acquired more than enough contentment for himself, he should overflow to others to increase more happiness for himself. Of course, the problem is to get yourself filled up and content first before you can overflow to others. Transcendental Meditation gives man that contentment one needs so he can overflow to others.

EGO poses resistance to drawing the Unified Field and its attributes to positivity and to fulfillment in life. The resistance to fulfillment is blocked by the EGO, the same as friction act as the resistance to the flow of electricity. This is my comparison to the energy that gives fulfillment in life.

Science stays away from biblical interpretations, but I can still refer to a similar biblical passage, "Whatever you do to the least of my brethren, you do it to me." Doing good for others eventually get support from the master you serve. The cause of natural law is to uplift the well being of humanity. So if you work for natural law and its cause, you gain spontaneous support of natural law.

That is why Maharishi says, "You become master of natural law. Natural law serves you and provides fulfillment for your desire."

Even Jesus Christ said in the Christian bible, "He that is the greatest among you, let him be your servant." Many theologians were baffled by his quote. Now I can relate this quote with the wisdom expressed by Maharishi's Ego Chart.

The more you serve the cause for others the greater power of nature is with you. The power you can posses is the greatest of all power because the Unified Field is the most supreme of all energy in nature. I began to understand Maharishi's instruction, "Do it. Just do it," he said. It is because nature's support will follow you. This was what Maharishi meant. That was why he would always repeat each time. "In honor of my master Guru Dev," not in his honor.

Even in his deathbed Maharishi said, "Now, I have completed my service to Guru Dev." All of the fifty years of Maharishi's life, he was doing the cause of his master, Guru Dev. It was the cause for the transformation of the whole world. Serving the well being of man is the goal of natural law.

That was why Maharishi always got support of natural law. When Maharishi needed a home, someone offered a home to him, just like that. The greater support of nature comes with greater service to others. Like the above explanation of electricity, the light is brighter when another light bulb is lighted. The more light bulbs lighted up, the greater is the brilliance.

The illumination is exponential in nature, not one plus one equals two, linear equation, but two squared, equals four, four times four equals sixteen and so on, five square, etc... like that.

Exponential increase is the character of energy fields, electrical, magnetic or gravitational field. The Unified Field has the greatest exponential power of them all. If the Ego does not exist above the Unified Field diagram, fulfillment in life becomes exponential too.

Maharishi's formula is the square root of one percent of the population practicing TM and Yogic flying will provide the exponential invincible effect. The power of group of minds at the level of silence, at the level of the Unified Field is exponential, similarly held by all energy fields in nature. The electrical energy is exponential to the square of the electrical particles within the bulb with brightness increasing correspondingly as the number of bulbs is increased. With service to more people, the EGO slowly fade away, greater support of nature will be prevail.

The Unified Field within the mind of man is an energy field, as described in Maharishi's Technology of the Unified Field. In man, Ego blocks natural law and support of nature. Like a tree, if a stone blocks the root, the root cannot draw the water from the soil. In man, Ego blocks the drawing of the power from the root energy field that gives fulfillment in life, beyond suffering. Without the blockage at the roots, the tree can draw water and spontaneously grow and bears fruit. It is a natural process of nature.

Without the Ego, man can draw the root energy field at the silent level of his mind and spontaneously live in fullness, free from suffering and problems. By Maharishi's precepts, this too, is a natural process of nature for man to draw from. Without the Ego blocking the root of the energy field, Maharishi called the Unified Field, life becomes problem-free.

With constant and consistent practice of Transcendental Meditation twice a day morning and evening, eventually God consciousness becomes a permanent state occurring in time, then support of nature will follow too. Transcendental consciousness evolves to God consciousness, by then the Ego has faded away.

Maharishi has described God consciousness when all cause becomes the cause of service to others and no longer directed towards the self, then the Ego completely fades away and nature comes spontaneously and provides support for your desire. This is the state of fulfillment of life when the Ego has faded away, all intent and purpose become in tuned with natural law.

The mechanics of the Ego chart affect world view thru the actions of leaders who rule each nation in the world. Nations that are at peace and whose goal is the expansion of peace and progress for everyone, they are in tuned with natural law.

Snarled Ego tendencies in human behavior are believed to be the cause of conflicts and military encounters. Scientific researches show Transcendental Meditation repairs snarled human tendencies and Yogic flying unify diversified ideologies that cause conflicts.

Albert Einstein said of the acceptance of new discoveries, "There are three stages, first, they ridicule, second they oppose. Third, you see the next generation using it." Today, it looks the reverse - it is ridiculous seeing many nations at war and in poverty, when a technology to remove them is already available with the application of TM and Yogic flying.

The Ego chart locates the Ego beyond the mind and intellect, it blocks the Unified Field. Without the Ego, the Unified Field is directly accessible, this means that support of nature will always be within reach and fulfillment is available all the time. The Unified Field is the field of fulfillment. It is the field of the total potential of natural law. The Ego blocks support of natural law. Like in the electric field, friction resists the flow of electricity within the copper wire acting as the conductor the electricity.

The elementary particles of nature display a character observed in quantum physics, where fine particles were found to create a field, like electric field, magnetic field or gravitational field. The finer particles they become the more invincible field they produce. The human mind at the level of silence creates a field, the

Unified Field. The human mind and nervous system act as the conductor of the Unified Field. Like friction resists electricity, the Ego resist support of nature and blocks the Unified Field. Support of nature means fulfillment of desire.

Serving the well-being of humanity is the cause of natural law. Transcendental Meditation allows the human mind to identify itself with the Unified Field in pure consciousness. The Unified Field is the seed form of natural law.

Throughout history, those who displayed aggression perished in disgrace like Hitler. Japan's aggression at Pearl Harbor led millions of its citizens to perish in Nagasaki and Hiroshima from the atomic bomb blasts in 1944. As the saying goes, "As you sow, so shall you reap" or "If you live by the sword you perish by the sword." Natural law is the most powerful inevitable force in the whole universe and violations of natural law incur a proportionate resultant reaction. Natural law is precise and final, like at 100 degrees centigrade, water boils, no more, no less and no excuses in between.

The United Nations granted the West to strike Afghanistan in October 2001 and Iraq in March 2003. Nearly half a million people were killed, less than one third of them were innocent civilians. Killing innocent civilians is against natural law. The 2008 recession could have been the resultant reaction. It affected almost many nations in the world, probably the nations that voted for the wars. The economic impact was likened to the status of the world after World War 11. The recession caused many homes foreclosed, thousands of lost jobs, impoverished many and indebted the USA to a degree equal to the world's combined debt. Natural law does not discriminate small or big country. Natural law always reacts to any action of force performed by anyone from any country.

In the Ukraine war in 2014, it caused nearly 7,000 killed, with Russia providing arms and supporting separatists in Eastern Ukraine. Russia is currently experiencing recession. With natural law, the aggressor is always the loser.

Depending on the number of innocent people killed, natural law react accordingly. There are no accurate number of victims from Russia's bombing of Syria, but the recession in Russia shows how many are becoming poor probably as many killed in Syria. Any action that kills is an action against natural law. Leaders hold the destiny of the people by the decisions they make. Leaders who declare war will transfer the effect to its own people, which will impoverish millions in the population. War kills people, natural law will react.

There are phrases in the Christian bible that says, "If one throws you a stone, throw back bread." Natural law supports the cause for the betterment of people and there is no excuse whatsoever to be exempted from the resultant cause and effect. "Revenge is sweet, said the Lord, but it is mine." Revenge is not for men to exercise, as it says, it belongs only to God.

Maharishi had said that the laws of nature are the will of God. The decree of natural law is inviolable and omnipotent. It is like water when it reaches its boiling point, it boils. When the human mind has fully accessed the Unified Field at the level of silence in large groups together, problem and suffering disappear. It is the decree from natural law!

Most conflicts are encounters of the Ego. In the civil war in Syria, President Obama of the USA intended to bomb Syria for President Assad's gazing innocent children and civilians. Then later, he changed his stand, instead he sent food and medicines. Opponents criticized him and called him weak. Resisting killing shows strength not weakness.

In the Ukraine conflict, President Obama did the same, he sent food, medicines and body armors, instead of arms or a contingent of military that many in the opposing party insisted be sent. Many critics called the President weak. I call him strong, on the basis of Ego chart. It takes more strength to say no to conflicts and no to military encounter. His decision was based on the concern of people not being harmed, unconcerned about the criticisms of his opponents.

It is a great blow to the Ego, to call the American President weak considering the USA has the mightiest military force around in the world. He did not use the military might against Syria. Instead, he exemplified strength, by exercising restrain, it mattered to him that no one is harmed on both sides. Not using the greatest military in the world at ones command is an exercise of strength. Any weapon that kill is an arm of weakness. Restrain in the use of arms, is a sign of strength. Greatness and strength comes from serving, from being a servant! Show of force is a sign of weakness, not strenght! How strange!

The use of force to expand influence is a sign of weakness and the Ego expanding. Service and kindness creates larger sphere of influence and gain more support of nature. Strength is embodied in the laws of nature beyond the Ego at the level of the Unified Field which holds exponentially more powerful effect. Aggression is a sign of weakness, not strength, no matter what pride and victory the winning party enjoys, natural law will soon seek its own irreversible course. Natural law rules by cause and effect with no exemptions. Violations of natural law has an equivalent impact and effect. Leading and living in accord with natural law is the basis of strength. Killing people is a violation of natural law. This is mine, is not good. This is ours, it is better. If nations help each other, exchanging food and medicine instead of exchanging bombs and missiles, peace and prosperity in the world will prevail. The Ego Chart truly illustrates the science of fulfillment and happiness for mankind. The Ego Chart determines who is the winner and who is the loser.

Invasions, conflicts or the use of military force are signals of expanded Ego. Arms should be used only to protect people, not kill them. International law is violated when arms are used to invade another country. It is also a violation of natural law. Expansion of boundaries thru force is expansion of the Ego. This is my interpretation of the real meaning of the EGO Chart. It should give a new meaning to world views. It should define who is weak and who is strong. It should guide decisions of world leaders. Their decisions may unknowingly harm or nourish their citizens. The Ego chart should guide decisions of those that hold the destiny of people.

To me, it is the most significant illustration of gaining happiness and fulfillment for every citizen. All leaders of nations should learn the wisdom and knowledge behind Maharishi's Ego Chart. Service to other nations in need illustrates exceptional character of a nation. The USA came to the aid of the Philippines during the super typhoon Haiyan. The USA sent a navy fleet to help locate the Malaysian Airline 370 that perished in the Indian Ocean and it send a team in Nigeria to locate the kidnapped Nigerian girls. The USA showed intent to support other nations in need.

Nations that provide aid and support in times of trouble are of exceptional characters. Exceptional character is displayed with service that benefits others. Service cultures the heart and obliterates the Ego. Soon as the Ego fades away, support of natural law flows unconditionally and the gates of happiness and fulfillment opens. It is available at the level of the Unified Field located at the most silent level of the human mind. This is the significance of Transcendental Meditation, it quiets the mind in an easy effortless manner.

The expansion of freedom, happiness and prosperity is the purpose of natural law. No weakness can exist at the level of the Unified Field, just like no impurities can exist in the magnetic field. The Unified Field holds the formula which brings life to true greatness beyond problems and suffering.

The Unified Field of modern science should be explored and embraced by governments instead of sticking to old histories. The old theory that the earth is the center of our universe that was known and prevailed throughout the world in the sixteenth century, it was replaced by the sun as the center of the universe. Galileo changed the old belief that the earth is flat. He found the earth round and orbits around the sun. The discovery of telescope proved Galileo's theory is true. Now the International Space Station overlooks the earth moving around the sun.

Maharishi has changed the old history of war and conquest. Maharishi has found the application of the Unified Field as the superior tool to tackle conflicts and war without damaged and that is effective. Bombs, missiles and military arm has become an arm of the ignorant past. The Unified Field has surpassed the use of atomic and nuclear arms that cause harm and cost enormous billions of dollars. The Unified Field is simple and easy to form, just by gathering large groups of TM and Yogic flying together, and allow natural law to take over and take its natural course.

Maharishi's interpretation of the ancient Vedic literatures led the way for modern science to discover the Unified Field. In the ancient Veda, it says, "In the vicinity of Yoga, no negativity can exist." In the scientific format of the Unified Field, like in the magnetic field, more magnets joined together creates a stronger and greater invincible magnetic field where no impurities can exist.

In Patangili's Yoga Sutras it says, "Yogic flying unify diversified influence in the area." The Unified Field proved to be the technology to remove problems and conflicts. Leaders of nation should take a parental role in finding solutions to its problems. Like a parent with a sick child, they explore all options to find the cure for their child. No mother or father will ignore any option to find a remedy for their child who is sick, no matter how unconventional it may seem. Yogic flying seems unconventional and originating from an ancient Vedic culture that the Ego will be tempted to say, "We are in the modern times, what can an ancient norm do to solve modern problems of the world?"

Scientific evidences already exist showing that the snarled tendency of the human behavior for confrontation and conflict can be repaired by the practice of TM and Yogic flying. They originated from the ancient Vedic literatures and the science brought about by the discovery of the Unified Field occurring at the silent level of the human mind now confirmed it to be true, it can remove conflicts that cause war and suffering

We have seen now, there are conflicts arising here and there, in the individual, in governments and in the world. There were no remedy effective enough for the attainment of permanent world peace. Today we see so many conflicts in Ukraine, Iraq, Libya, Syria, Yemen, Afghanistan, Israel and Palestine who have been fighting since time immemorial since biblical time. No amount of peace treaties were able to create permanent peace for them.

Many parts of the world are in turmoil and needs to be solved. Countries confront many killings from lone wolves. Other parts are struggling from storms and hurricanes, others from droughts and fires. Middle East countries are struggling from terror and war! In God consciousness, man becomes conduit of natural law. The Unified Field serves the cause of natural law. It can balance climates, create peace and prosperity. Maharishi said that natural law is the will of God. The cause of natural law is to serve and nourish man, not to destroy him. Maharishi has brought to light the capability of the human mind to prevent any negative trends from occurring, at the level of silence, in large groups together at the level of the Unified Field of natural law expressed by modern science.

For many centuries, great men have spoken about this great wisdom within man, the God within. Even Jesus Christ said, "Seek ye first the kingdom of God within and all these things will come unto you." (Matthew 6:33). "The kingdom of heaven is within you." (Luke 17: 21). Plato and Aristotle, who were great thinkers and philosophers of previous centuries, they spoke of the inner silence within man.

St Augustine said, "I entered into the innermost part of myself." St. Gregory the Great, an heir to a large Roman family fortune decided to become a monk and became a Pope, he wrote, "The mind is frequently carried away into the sweetness of heavenly contemplation - it sees something of the inmost realities as if it were through the mist."

St. Theresa of Avila wrote, "My soul at once became recollected and I enter the state of quiet rapture." Thomas Merton, a learned graduate from Columbia University and a famous American monk wrote, "The utter simplicity and awareness of the infused light pours into our soul suddenly over comes us to a new level of awareness."

Most of them have enjoyed contemplative silence, not like living in our most modern life with TV's, internet and modern gadgets that keep the mind engrossed in the fast pace of modern living. Silence now seems to be unimportant and seldom available.

One need not become cloistered. In the fast pace of modern life, Transcendental Meditation can bring the silence that anyone can enjoy any place, anywhere. The path to God is easy. God consciousness is made available to all religions. In Buddhism, it is called Nirvana. In Islam, it is Fana. In the Indian mythology, it is called Yoga. Maharishi, Einstein, and Hawking call it the Unified Field of natural law.

Philosophers, theologians and scientists have spoken of the same state of transcendence experienced in the inner most sanctum of the self. I call it the field of the anima solo particles, the finest most powerful particle of nature, the unseen "sap" within every creation. Call it by any name it is the same banana, ancient, past, and present. The future of science will produce a name, but it will still be the same banana describing the qualities and attributes drawn at the quiet level of the mind.

Most scientists do not believe in God. Maharishi as a scientist and physicist expounded God consciousness as the evolutionary path to God and believed that there is a God that created order and design in the universe. Maharishi wanted all men to experience the will of God easily thru the practice of the Transcendental Meditation technique that he brought out into the world from the deep forest of the Himalayas.

Isaac Newton believed that the solar system could not have been created without a God. Stephen Hawking believes there is no God and challenged Newton's theory of God and referenced God as a force (or maybe the force field described in Maharishi's Unified Field) that could explain the creation of the universe. Stephen Hawking stated "That God is defined as the embodiment of the laws of nature and because there is a law such as gravity, the universe can and will create itself from nothing, spontaneous creation is the reason there is something rather than nothing. That is why the universe exists, why we exists. It is not necessary to invoke a God to set the universe going."

My understanding of spontaneous creation is that there is something unseen and unnamed that contains the most powerful seed form of nature occurring in the universe that creates "from itself, by itself, within itself" just as Maharishi has said. It can be drawn by the human mind in silence, at the level of the Unified Field.

From the seed within itself, by itself alone and by its own process, it creates the diversity of the universe. The mind is the seat of the total potential of natural law, of the Unified Field. It creates from within itself, by itself alone from nothing and lingers at the most silent level of the human mind. You live in accord with the will of God or the field of the Creator of the universe or the field where the grace of God exists.

I believe the theory of Stephen Hawking that "science will eventually win because it works." I believe Maharishi's theory of the Unified Field occurring in the silent level of the human mind. In time, scientists will be able to identify the name and attributes of the solo particle, the root or origin of all the elementary particles of nature, mean time, I will call it the ASu particle, the anima solo particle that nourishes and animates man and the whole universe.

16

GOVERNMENT ADMINISTRATION THRU NATURAL LAW

"The world has witnessed the rise and fall of monarchy, the rise and fall of dictatorship, the rise and fall of feudalism, the rise and fall of communism and the rise in democracy, and now we are witnessing the fall of democracy, the GREAT FALL now will never rise to fall again. The river of time flows, the theme of evolution of life continues, sweeping away with it all that does not blossom into perfection. Now with the fall of all man-made systems of administration, a new quality of administration is dawning based on the administration of nature's government-perfect, eternal, and true for all time." ---*Maharishi Mahesh Yogi*

Maharishi spoke about the inadequacy and complexity of man-made governments. Looking back thru history, many leaders have been unable to satisfy the well being of its citizens.

In ancient Rome, Julius Caesar was assassinated by a disgruntled party. In Russia, the ruling Czars eventually faded away thrown out by its own people. In Germany in the 1930's, Hitler was not hailed by all of its citizen. A military conspiracy plotted his assassination and then disgraced in battle, Hitler killed himself.

In my time, during President Ferdinand Marcos era in the Philippines, the people revolted against his rule and eventually fell out of power. In the recent uprising in the Middle East, the same thing happened in Egypt with Hosni Maburak and Mohammed Morsi. In Libya, Muammar Kaddafi was dragged in the streets and killed by mobs against his regime. With all his gold and grandeur, the mobs in the streets pulverized his grand palace to the ground.

The United Nations is unable to quell the uprising of the people of Syria against President Bashar al-Assad. Meanwhile, thousands of people are killed and millions are dislocated as refugees. mostly women and children living in severe conditions along bordering countries. Many risk fleeing Syria to Europe and drowning at sea. Rebels have gathered with the intent to remove Bashar al-Assad in power. Millions of the population are fleeing the regime. Russia, Syria's only ally is trying to prop him from falling out of power by bombing rebels against Assad, claiming also barely bombing ISIS.

Few leaders can lead satisfactorily and be truly acceptable by its own people. When economic and political agenda becomes controversial, mobs rise and terror groups are formed that topple leaders like the uprising in Ukraine against former President Victor Yokunovych who fled to Russia fearing for his life.

No leader can be sure he will be in power long. Many leaders in history came into power and later placed out of power by the very same citizens who voted for them. In the United States history, Abraham Lincoln was assassinated. John F. Kennedy and his brother Robert, they were shot too. Currently, in the USA, the economic gap widens and the great divide is increasing among the one percent and ninety nine percent in the general population. The reduction of the middle class has greatly expanded due to the 2008 recession. The trillions of dollars in government debt slowed down economic growth and has become a controversial political agenda.

The increasing divide between Wall Street and Main Street is affecting the world's greatest democracy. People expect one man, the President to be able to provide prosperity to more than three hundred million citizens! Never was there a US President, like Barack Obama whose social programs were so publicly opposed, contradicted and despised like the Affordable Care Act project that caused so much controversy, nearly shutting down the government. Wonder why a project that benefits millions of citizens created so much conflict brought by partisan politics!

That is why Maharishi called on governments to validate and utilize the Unified Field. It is the government of nature that can eliminate problems otherwise conflicts and problems will continue to occur. Maharishi described how the Yogic flying program can be successfully be used which was described in the ancient Vedic literatures, it states, "In the vicinity of the phenomenon of Yoga, Yogic Flying, conflicting tendencies disappear. The unifying influence neutralizes the diversifying qualities in the area."

With man-governments, problems and conflicts will occur and repeat the same cycle of conflicts and oppositions. Peace will never be permanent. Warring parties will always occur as it has been repeated for so many centuries. The leader or President of any country whose intention is pure in the service of its people, he should be the one who will implement the technology of the Unified Field so that his people will be blessed by a better economic life and provide an effective programs for its citizens. There will be a more orderly and peaceful government administration and leading the country will be with least effort and easy. The Unified Field of natural law will provide the strategy for a permanent transformation of society for every country.

The President of any country who will utilize the knowledge of natural law with this simple natural phenomenon and use this system of governance of nature, he will be exalted among all other leaders of the world. His country will be prosperous and economically self-sufficient. His citizens will be healthy. His country will be crime free. Its citizens will become fully enlightened individuals. All aspects of governance will be in accord with a natural law that is problem free. Otherwise, any leader who comes in with pure intent will later realize that his efforts will stay unappreciated. He will become exasperated with the magnitude of problems and he will eventually experience the difficulty of running a government. He will be constantly attacked, like many world leaders who have been put out of power, they cannot govern effectively no matter how good their intentions are.

Just like all the leaders in the past who have tried their best efforts and failed to satisfy the needs of the people. Soon as the people start to protest, the leader starts to lose power. There were many leaders who have been placed out of power in modern times just as Kings and Emperors dethrone in ancient times. The same story repeated over and over. Anyone who would like to lead people or nations, should consider Maharishi's absolute theory of government, the government in accord with natural law that is problem - free.

Natural law based government is a government run automatically bringing prosperity, peace and harmony to the whole nation. Like a tree, just water the roots, nature takes it course. TM and Yogic flying understandably looks unconventional but it works! It has worked in Mozambique, Africa when then former President Joaquim Chisano had his military did Maharishi's program in 1994.

The President of Mozambique, Africa validated the TM and the Yogic flying program. President Chisano of Mozambique in Africa learned TM in 1992. His country was at war and drought prevailed. People were in poverty. When they were presented the use of Maharishi's technology of the Unified Field, they had the option to try it instead of bearing years of war, poverty and drought.

Soon as the military implemented Maharishi's programs, the predicted results happened. War stopped. Rain fell in drought areas. Economy rose 19%. New found of coal, natural gas and petroleum provided thousands of employment and these new discoveries raised the economic index of the Mozambique nation. Mozambique rose from the poorest country to the fastest growing country in Africa from 1993 to 2004.

President Chisano's tenure in Mozambique utilized Maharishi's program of TM and Yogic flying that laid out a government in accord with nature that is efficient and created economic prosperity.

Present man-made government cannot guarantee it can provide the fulfillment of the needs and goals of the majority of its citizens. People look up to the government for support, but support is seldom available. People can now look and go inside their mind and find a perfect natural law based government to support and provide means for their well being. Just by creating a large groups together equal to the square root of one percent of the population of the country, automation in government administration will prevail.

It is the government of nature within that works and provides support. Every leader now has a tool to stay in power and be a witness to that perfect and spectacular rule of nature that satisfies the fulfillment, prosperity and freedom of its people. The government of nature will rule in harmony and order. Then, there will be no need for conflicts, protests and oppositions. The order within man will parallel the order and harmony of the functioning order of the universe. Utilizing the organizing power and intelligence of natural law is the best option to lead. This is the only option that will guarantee a perfect government satisfying the needs and goals of every citizen.

The Unified Field of natural law is the only safe and effective endeavor a leader can provide for his people. Then a life of happiness and fulfillment will prevail among the citizens. This endeavor could be easily implemented by any leader of any country because already in a country, there are groups of military in place or police staff exists and ready to defend the country and provide order in society. All they have to do is form a group together and participate in the TM and Yogic flying program together twice a day and spontaneously order will prevail. Prosperity will flourish. The application of the Unified Field found enlivened within the human mind at the level of silence, in large groups together, it will produce a spontaneous super radiance of positive trends prevailing. Just like small magnets joined together create a bigger magnetic field and corresponding exponentially bigger magnetic force field invincible to outward negative trends. The Unified Field provides an invincible shield and as soon as it spreads over the country or the world, immediately negativities will fade away, only positive trends will prevail.

The USA with the greatest and most powerful military, the external terrorist attack of 9/11 was not prevented. Home-grown terrorists exploding home-made bombs and shooting guns of were not prevented either and constant killings happen even inside military stations. With all the military expertise made, negative trends cannot be prevented. Maharishi's absolute theory of government will make any country invincible from any negativity and free from any attacks. This theory applies to any government either democracy, communism, or even dictatorship. Natural law government does not discriminate, small or big nation, it doesn't matter. What matters is, each country will be self – sufficient, at peace and prosperous. Invincibility of every country will be guided by the laws of nature. Just like no impurity can exist in the magnetic field. No impurity occurs when liquid water turns to ice crystals, at zero degrees. Invincibility in man occurs at zero excitation of the human mind in large groups together.

When the required number of citizens practice Transcendental Meditation together and reached zero excitation of the human mind on a permanent basis, no negativity in the country will prevail. The country will have just sufficient number of military to maintain defense, no more no less. No soldier's life will be wasted. No woman will be widowed by war. No young children will become orphans. Police squads will become peace squads. The military will become defense peace keepers.

Wars will become obsolete events of the ignorant, primitive past. Wars will not happen anymore. The groups of TM and Yogic flying participants will be the peace and prosperity teams. Homeland security will become the silent mind security. Four-billion-dollar bomber planes used in wars will no more be needed. Defense budgets running trillions of dollars will be used for education and social services. A phase transition will automatically occur from negativity to positivity for everyone with least effort. The government of nature will prevail. Heaven on earth will become a reality.

Maharishi postulated the formula and the number required for the phase transition to occur. It has been validated by several scientists. It can be validated again with similar results. Maharishi was a physicist, a scientist himself. He came up with the Maharishi Effect formula.

Pythagoras postulated the mathematical formula of a triangle. Pythagoras and Maharishi postulated a theory. The proof is in the process of verification. If the verification is validated to be correct it becomes a fact. Pythagoras theory was verified and validated. Likewise Maharishi's theory of the Unified field has been validated by more than 600 scientific papers from world renowned universities worldwide and the group effect was repeated and validated more than fifty times. That is why Maharishi has invited governments to validate his Unified Field theory and the results will be a surprised.

Any big nation who will establish the 9,000 for the world will expand its worldwide influence of prosperity and peace. Wars will end. Conflicts will be resolved. Economies will rise. New markets will open. Climates will balance and all negativities will fade away.

The Maharishi Effect will change the history of the world from suffering and conflict known for many generations, now prosperity and peace can be easily attained with the discovery of the Unified Field. Heaven on earth is the next history of mankind. The administration of natural law will prevail in every government that will be free from conflicts. With the group established, no opponent will rise. No critics will undermine government programs. No enemy will be born. Many peace treaties were signed thru the United Nations, but many more conflicts rose more than the peace treaties that were signed throughout the existence of the United Nation during the past decades. Conflicts and wars come and go and the world never came to a permanent peace.

Enormous amount of money, trillions of dollars were spent by the United States in the two wars, in Afghanistan and in Iraq. One fighter bomber costs nearly four billion dollars, it did not create permanent peace in Iraq. After nearly ten years since 2003, another conflict in Iraq has risen. The large amount of money previously spent in Iraq lay wasted and in vain. Recently ISIS militants have risen and have taken over many cities in Syria and in Iraq.

For more than one thousand five hundred years, the conflicting groups in Iraq, the Shiites, Sunnis, Kurds, Hesbollah and other non- denominated groups have been fighting for control over its vast oil reserves and the authority to rule. None of the groups want to be dominated by any of the other groups either. No unifying principle was ever applied to unify them that can make conflicts cease. Maharishi's natural law principle can end their diversified conflicts that has been going on and off and on again for one thousand five hundred years. It is too much suffering lasting that long!

In Iraq and Middle East, the USA suggested a the ruling government that include sects, but without Yogic flying in place, conflicts among all or any of the groups will not end soon or if it does, sometime in the future they will fight again and the conflicts will repeat again, like it has since biblical times. The verses in the Veda states, "Yogic flying will unify diversified influence and tendencies around the area." Several events have illustrated that it does work. After years of conflict and the trillions of dollars spent by the US in 2003 until it pulled its troops in 2011, they have all gone to waste. It is about time Maharishi technology of the Unified Field be used on a permanent basis so that peace and prosperity will prevail in the Middle East or in any other nations at war. It is the same with Israel and Palestine, they have been fighting since biblical times. Nothing has come out to bring them permanent peace even during our modern times today. Israel, Palestine, Iraq, Syria, Egypt and most Middle Eastern countries were supposed to be among the oldest civilization and should have evolved better, but seems like evolution to become better governed have reversed, because they have taken more deadly modern missiles and arms that kill masses of people instead of the old bows and arrows used thousands of years ago in ancient times. Military arms will not end the conflicts from recurring again, maybe for a while, but what about another ten years from now or twenty years? Resolution of the conflicts will cost more than ignoring the use of Maharishi invincible defense technology, the Maharishi Effect that can provide permanent peace at so little cost. Governments are the beneficiary of the Maharishi Effect with little cost to expand peace and prosperity for all nations.

Maharishi at one time, called for governments to provide one billion dollars to support his project creating 9,000 Vedic pundits boys in India, which group intent was to create permanent peace for the whole world, no one single government gave him a single dollar.

Instead, Maharishi gathered private donors from peace-loving individuals practicing Transcendental Meditation and used it to pursue his Yogic flying Vedic pundit boys persistently to the very last days of his life. Peace and happiness was Maharishi's sole purpose for mankind. Governments cannot anymore ignore Maharishi's Yogic flying because there are several existing evidences that show it works. Maharishi's Yogic flying Vedic pundits boys are still going on today at the central part of India, sustaining barely one thousand Vedic pundits boys today. A group of 9,000 is needed for permanent world peace to occur. Maharishi's purpose was to create an environment where no mother will lose a son or lose a husband fighting a war, or allow mothers and her children live miserably in refugee camps due to war. With the number required for a country established, no leader will be toppled down by militants or allow any nation be isolated from the world. That is why Maharishi had repeatedly called on governments to apply his invincible defense technology to end conflicts, otherwise conflicts will rise again and again like it has in the past and are now constantly occurring again among some nations at war.

17

THE GLOBAL COUNTRY OF WORLD PEACE

Tony Nader, MD, PhD, Maharaja Adhiraj Rajaraam

Maharishi founded the Global Country of World Peace, a country without borders that supports the sovereignty of every nation and that which upholds the principle of silent administration of natural law enlivening positive trends in every county throughout the whole world.

It was established in October 2000 as the global foundation to perpetuate Maharishi's teachings interpreted by modern science originating from the ancient Vedic literature of India. Maharishi Vedic City in Fairfield, Iowa, USA is the Capital of the foundation.

Because of his significant discoveries, Tony Nader, MD, PhD was crowned first ruler of the Global Country of Word Peace. Maharishi gifted him with his weight in gold and named him his successor. Maharishi gave him the supreme title of Maharaja Adhiraj Rajaraam.

Many organizations in the world have established groups whose functions are to create peace. Permanent peace cannot be obtained at the manifest level of creation. Like a plant, if the leaves are decaying, water the root, the unmanifest and unseen part of the tree. Watering the manifested leaves does not solve the problems of the tree. Watering the root nourishes the whole tree.

The significance of the Vedic technologies refers to the unmanifest, non-material aspect of man at the level of silence that can be explored thru the practice of Transcendental Meditation. The mind is the root and the seed form of natural law brought to light by the recent discovery of the Unified Field.

Many problems rise in society and several means were tried and used by responsible leaders, but most does not work on a permanent basis. Well- intentioned leaders with high ideals have led their nations to the illusive path of prosperity and peace. Many have failed and many will fail again in the future unless the root of life lingering within the silent level of the human mind becomes the main target of recourse. Only a technology based on the non-material, unmanifest level of nature will create permanent peace and prosperity. Maharishi's Vedic technologies which the Global Country of World Peace hold has been interpreted by modern science to effectively create permanent peace and prosperity. The invincibility of the Unified Field of modern science has the natural character to permanently uplift life to a status of heaven on earth.

Because like the farmer who waters the leaves which are decaying, instead of the root, it that does not work. Watering the root governing man has its basis in Vedic science that is now interpreted by modern science at the level of the recently discovered Unified Field of natural law. The Unified Field describes the unmanifest seed form of nature occurring within the silent level of the human mind.

The basis of diseases of societies are found within the human mind. It was defined in the preamble of the UNESCO Constitution which states, "Since wars begin in the minds of men, it is in the minds of men that the defenses of peace must be constructed."

The Unified Field describes the level of the root of man and must be the working flat form where leaders should take a stand. Since the root is the basis of the leaves, it must be the target of attention. The target of attention in the transformation of man and society must be the study, understanding and application of the Unified Field.

Einstein located the atomic level that led to the development of the atomic bomb that ended World War II, but other wars did not end there, it continues until now, there are wars still raging in the world. The Unified Field contains the finest of all particles of creation further beyond nuclear and atomic form, that is why it is exponentially powerful. It is a natural phenomenon that finer levels in nature increase its field effect exponentially and proportional to the number of particles concentrated in form within its field.

Any nation at war has now a technology to end it and prevent it from occurring again. By having a group of square root of one percent of the nation's population practicing the TM and the Yogic flying program together in one place, it can neutralize the enmity between parties. Rebels vanish. All pent-up anger in the collective consciousness of nations at war will be diffused. A general peaceful and prosperous environment will prevail. Not only peace and prosperity prevail, climate change and global warming will be balanced. Drought, fires, floods and hurricanes that constantly occur on earth will fade away. The Unified Field displays a more powerful effect produced by large groups participating in Maharishi's programs at the level of the finest particles of nature, the AS_U particles, described in the science of particles on Chapter 7, in contrast to the

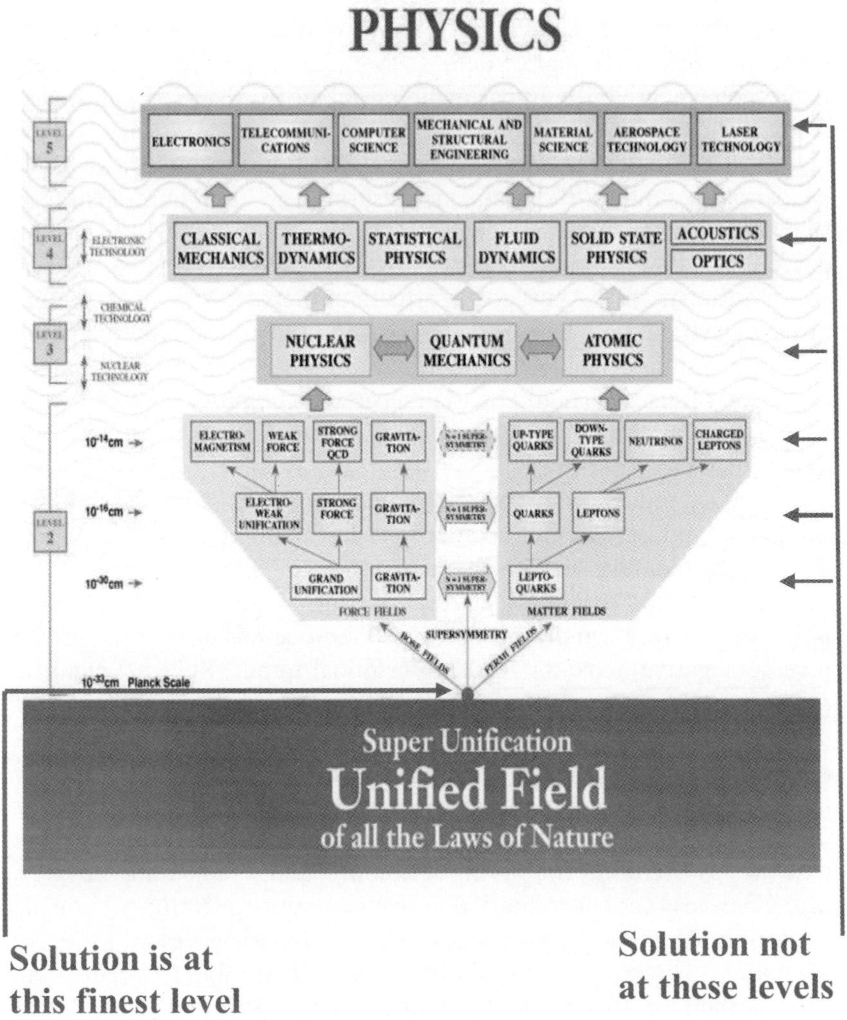

THE UNIFIED FIELD OF NATURAL LAW[40]

[40] *Global Country of World Peace*

atomic effect made of concentrated hydrogen particles create destruction, because finer levels in creation are exponentially more powerful than the hydrogen atom, it creates nourishment and does not destroy. The ASu particle is the finest, finer than sub-atomic, nuclear and even beyond the leptons and quarks.

The goal of the Global Country of World Peace is to provide the technology of invincible defense that can support any government from their sovereign boundaries being taken over due to aggression from another nation.

The goal is to improve education thru consciousness-based education expanding the brain of every student. The introduction of Maharishi's Transcendental Meditation program expands the brain potential of every student.

Maharishi's Ayurveda is effective in preventing and reducing diseases. All public health servants should take pride in the reduction and prevention of disease instead of building new hospitals. The reduction of disease should be the metric of accomplishment. TM and Maharishi's approach to health have been verified to be effective in the prevention of diseases.

The reduction of prison population is a great burden that needs to be addressed, but no solution is available unless they turn to the Global Country of World Peace who are the custodian of the knowledge to easily and effectively transform man. Prison population is increasing in alarming proportion. Transcendental Meditation can reverse negative behaviors causing criminal trends. Several scientific research indicate improvement of intelligence and enhancement of positive behaviors.

Any nation faced with economic problems can utilize the technology of the Unified Field, the Maharishi Effect which is the most effective tool for progress. The Vedic economic principles out date any system of improving economy and eradicating poverty. Many nations race for military dominance with huge military budgets, yet within its borders poverty still exists. Maharishi describes Vedic economy, "Do less and accomplish more." It is like a fruit tree, so long as the tree is watered at the root, the tree will prosper and bear fruits without effort. Nature takes its course. With the technology of the Unified Field in place, in every country will be prosper. Nature will take its natural course. Just by establishing a square root of the country's population practicing TM and Yogic flying in groups together, determines how the tree of life in the country blossoms and bears fruit. It was repeatedly observed that when a square root of one percent of the nation's population practice the TM and TM-Sidhi program together in one place, economies rise. These occurrences on economy were illustrated by the Global Tate of Utopia event in Fairfield on Chapter 6 where stock markets rose all over the world.

A government that is not harmonious, plague by disputes, grid-locks is ineffective to produce programs of prosperity for its citizens. The technology of the Unified Field can provide unity of purpose and government programs will be administered supported by natural law enlivened within the quiet level of the human mind.

Easily success in government administration will be made available within the group of lawmakers because they are already in a group. Senators, congressmen and all their staff together will become a powerful force in creating a perfect government. All they need to do together at the same time is to try Maharishi's TM programs. Twice a day before and after work.

Air turbulence like tornadoes, storms, drought, floods and fires that frequent the country show nature's signals that problems are present in the surroundings. The solutions can be located within the human mind.

The Global Country of World Peace can layout blue prints for solving problems of every country. Ministries have been established in charge of every country in the world. It has scientists that can lay out a program to prevent problems before they arise or entirely remove the existing problems. The Global Country of World Peace has a track record of nearly fifty years and hundreds of validated experiments published in prestigious journals that indicate the effectiveness of these programs to transform man and his environment.

The Global Country of World Peace can provide an effective approach for developing individual and collective consciousness and can effectively target the root cause of regional, national, and global conflict arising. The Unified Field approach to national security and world peace was extensively field tested and backed by rigorous scientific research during the past fifty years of Maharishi's lifetime.

The Unified Field has basis of achieving a measurable effect of peace that a fertile environment for successful diplomatic and other conventional approaches can be created. It should therefore be the central component of any comprehensive approach to peace and national security. The responsible leaders of every nation and lovers of peace everywhere are warmly invited to utilize the knowledge kept by the Global Country of World Peace and achieve a new era of harmony for the whole world family.

Military groups practicing TM in South America[41]

Vedic Pundits in Central India practicing the TM-Sidhi program[42]

[41] *www.globalcountryofworldpeace.org*
[42] *Vedic pundits at the Brahmasthan of India*

The Global Country of World Peace in Fairfield Iowa, USA at the Maharishi Vedic City, is a proto-type city for reference. Their website is globalcountry.org. All the details of the Vedic programs are clearly defined. A new and effective approach to solve the many problems of the community, country and the world is at hand. There are still so much to be done. Permanent fulfillment of Maharishi's goals to create heaven on earth to solve problems of mankind is now available. There is no more need to suffer, a problem - free society is possible to be created. The applications of the Unified Field effect are in the hands of governments or people of prominence with means to create a group or provide the funds to support one.

Fairfield and the Maharishi Vedic City have been hosting groups for more than forty years. The foundation is there in place-in Fairfield. Increasing the number of participants on a permanent basis will guarantee invincibility prosperity for the USA.

The USA is the country that can easily implement Maharishi's programs because the county has the most TM practitioners with more than one million from the six million who have learned TM around the world. Many rich and prominent American citizens and celebrities have taken into TM which has gone mainstream in the USA as nearly 20,000 come to learn it every month.

TM and the Maharishi Ayurveda in Fairfield has also gone main stream in America. And because Fairfield is already thriving and facilities have been built, the USA has the greatest chance to create the exponentially powerful super-radiance effect on a permanent basis that can create heaven on earth.

For the world 9,000 participants in groups together are needed to create peace, prosperity and harmony in the world. The 9000 project at the central part of India is currently on-going with more than 1,000 participants. It is being supported by many American practitioners of Transcendental Meditation.

Maharishi in the same year that the Global Country of World Peace was established, he formed the Federation of Traditional Rulers of the World in October 24, 2000. All ministers of the Global Country of World Peace get the tile Raja, titles by the ancient rulers of the world. Traditional rulers were respected and honored by its people in contrast to two party system that compete and criticize opposite parties. With all the new technology of peace and prosperity that the Global County of World Peace has at hand, Maharishi has invited the traditional rulers to apply them in their nations.

Maharishi has said, "There are thousands of traditional rulers whom I consider to be the pride of all nations. These great traditional leaders have, throughout the ages, administered their kingdoms and territories with a great family love in parental role of administration. There have been very peaceful administrations on this level of great units of ancient civilizations and I welcome them to come together in the World Federation of Traditional kings."

"The Total Constitution of this Government of Natural Law is now available to us on this platform of Total Knowledge. It is my great joy to welcome all of you who are the units of cosmic administration on earth. Kings you are, holding the authority and power of Natural Law and maintaining orderly administration of society throughout the ages. That is your status, full of dignity, full of honor, full of love, full of harmony, full of peace for the whole of mankind in the long history of the world."

"My admiration for you is absolutely unlimited because you are representing that eternal tradition of administration of the universe - administration thru Natural law - which has been naturally handed down to generation to generation to generation."

"You are the representatives of infinity, eternity, freedom, invincibility. I welcome you all to re-create a tasteful world, to create a tasteful administration on all levels of administration."

"You have been beautifully honored in your units of administration of world peace. You are the glorious bright stars for the bright light of peace perpetual in the context of the infinite diversity of the universe. You deserve the honor."

"I welcome you with all the words of honor and dignity and harmony and peace that I have. I welcome you with great pride to adore on the same perpetual basis which your beautiful, glorious tradition of administration has been holding in the long history of the world."

-Maharishi

Maharishi crowned his successor, Tony Nader, MD, PhD, the Maharaja Adhiraj Rajaraam, King and First Ruler of the Global Country of World Peace. He now resides in Maharishi's mansion in Vlodrop, Holland, and rules in silence. Before Maharishi passed away, he assigned Ministers-in-charge of every country who serve and administer the global affairs of the Global Country of World Peace. Service cultures the heart and obliterates the ego paving a clear path for full support of nature described in Maharishi's Ego Chart. It is like Alladin and the tale of the genie and the magic lamp. The genie that serves Alladin creates the magic, not Alladin, the master. The Ministers that serve Maharaja Adhiraj Rajaraam will create the magic in the fulfillment of the goals of Maharishi's program to create heaven on earth.

After Maharishi passed away and in line with the great tradition of masters which perpetuates the continuity of the Vedic wisdom from the ancient Vedic civilization of India, Maharishi founded the Global Country of World Peace as the custodian of the holy tradition of masters. Its purpose is to keep pure the format of the ancient tradition and to maintain its effectiveness for the next generations to come.

The Global Country of World Peace has become the custodian of the Holy Tradition of Masters that originated from the ancient Vedic civilization. The Vedic settlers occupied the land along the river banks of the great Saraswati river extolled by the ancient Vedic literatures. It provided water and agricultural land for their crops to flourish, but thru the passage of time, the Saraswati river dried up and the Vedic settlers scattered and moved away searching for new lands. Soon after, the Vedic civilization faded away. Some ancient Rishis, masters of the ancient tradition settled in the Himalayas and there they kept and continued on the Vedic traditions. The tradition was transferred from master to student generation to generation. It was Maharishi's search for a master in the Himalayas in 1945, where he found Guru Dev, who took hold of the ancient knowledge of the Veda and Vedic literatures.

Fifty years since Maharishi came down from the Himalayas, the Vedic tradition of masters are extolled year after year, for bringing the great wisdom of fulfillment and happiness handed down generation to generation. Before Maharishi, Transcendental Meditation was the exclusive use of chosen students. Maharishi broke the tradition with the intent of bringing the ancient wisdom of the Vedic tradition to be enjoyed by everyone in the world and that all mankind should be the beneficiary of this great wisdom it held. The auspicious year of 1957 was celebrated as the inaugural year of Maharishi's worldwide Transcendental Meditation movement, but Maharishi has explained that the TM movement truly began in the blissful, infinite silence of his Master, Shri Guru Dev, Maha Yogiraj, His Divinity Brahmanand Saraswati, Maharaj, Jagatguru Bhagavan Shankaracharya of Jyotir Math, Badrikashram, in the Himalayas.

Guru Dev was revered as the embodiment of Vedic wisdom closest to Maharishi in the great tradition of Vedic masters who have passed down the eternal wisdom of life from teacher to student since time immemorial.

"The Vedic Tradition, upheld in its purity by a long history of custodians, enshrines the supreme knowledge of the integration of life. From time to time a revival of man's understanding of the eternal wisdom of this Holy Tradition arises to rescue him from suffering, restoring him to the speedy path of evolution, and awakening him to a meaningful life in fulfillment."

"The Masters of this Tradition have been exponents of reality from earliest ages. In each new epoch, they have propounded the enduring truths of practical living and have set out those standards by which men's lives may attain the highest achievements and fulfillment, generation after generation." *-Maharishi*

In the Upanishads, a Vedic ancient book says, "Who has a teacher, he masters the Vedas. Indeed, approach a guru for the sake of that treasure of true knowledge." This makes the wisdom of service to a master significant in seeking the totality of knowledge blossom in one's life. Service is the gateway to full support of nature. Service obliterates the Ego because one completely loose his identify and takes on the identity of the master he serves.

The wisdom of service nourishes the heart, then one gains support of nature because service obliterates the Ego, The act of service moves the attribute of love to the highest level. Service unfolds the totality of knowledge.

Maharishi's service to his master Guru Dev brought about the discovery of the Unified Field that brings Maharishi Effect, the supreme knowledge of life that will bring about heaven on earth. As soon as 2,500 participants of the Yogic program will be reach in the United States, all problems of the country will be neutralized.

When the threshold of 9,000 is reached in the world, the full blossoming of life on earth will be attained. The number is equal to the square root of one percent of the population of the world or each country.

That is why Maharishi is considered the greatest of all the masters for his discovery of the super-radiance effect exponentially increasing positivity in all aspects of life. The event in Fairfield confirmed the results of the Maharishi Effect in global scale done in 1983 shown in Chapter 6. This makes service the utmost honor a student can give to himself. It is the one who follows and serves that creates the magic, it benefits himself not the master he serves. This is the secret of the Vedic tradition of masters.

In the Upanishad, it said, "Who has a teacher, he masters the Vedas." Knowledge unfolds to one who serves a master. Maharishi's service to Guru Dev unfolded the Maharishi Effect and the Unified Field, the totality of knowledge of natural law underlying the quiet administration of natural law enlivened at the level of the human mind in large groups together. Even in his deathbed, Maharishi said, "I have completed my duty to Guru Dev."

The greatest service and honor a practitioner of TM and Yogic flying participant to Maharishi is to participate in the group program that benefits himself and his surroundings. Participating in the group program of TM and Yogic Flying is the greatest service that honors Maharishi, the great master, it benefits and honors man more, not Maharishi, he is gone.

In Fairfield, Iowa, those who participate in the group program at the Dome give the highest honor to themselves, to their family and to their environment. What greater honor is there than to support the cause of natural law and the creation of heaven on earth. It is not the master that creates the magic but the follower who serves. It is the practitioners of TM that will create heaven on earth. That is why service is the greatest honor one can give to a master that benefits the self more. Those who serve are the creators of heaven on earth. Thcy end wars. They create peace. They create prosperity for themselves, their family and the environment. All that is needed to do is to go and encourage others to participate in the TM program together with all the others. Soon as the threshold number of participants is reached, they can posses that power to create heaven on earth. Participating in the TM program is truly the greatest service to Maharishi.

This is the secret of the Vedic tradition of masters - the secret that those that serve creates the magic and has the power to transform the world. Just like Maharishi's lifetime of service to Guru Dev made him discover the Maharishi Effect that can transform the world. It was in 1977 that Maharishi recognized the group effect of the practice of Transcendental Meditation, and the Yogic flying by then Guru Dev has been long gone more than twenty years.

The Unified Field is an energy field like electricity, where the copper wire is the channel of electron particles in electricity, likewise, without the Ego, man becomes the perfect conduit of the Unified Field providing life in happiness and fulfillment. The Ego blocks fulfillment.

The Unified Field is the field of the total potential of natural law, total knowledge of natural law vibrating within the human physiology. The ancient Veda and Vedic literatures, brought to light the ancient secret knowledge of totality of life that was held by the Masters of the Vedic Tradition.

The following were recent masters: Narayana, Padmabhava, Vasistha, Shakti, Parashara, Vyasa, Shuka, Gaudapada, Govinda, Shankara, Padmapada, Hastamalaka, Trotaka, Varttikakara, Guru Dev (see the tradition of masters in the following pages) and Maharishi Mahesh Yogi.

"The Vedic Tradition, upheld in its purity by a long history of custodians, enshrines the supreme knowledge of the integration of life. From time to time, a revival of man's understanding of the eternal wisdom of the Holy Tradition arises to rescue him from suffering, restoring him to the speedy path of evolution, and awakening him to a meaningful life in fulfillment."

"The Masters of this Tradition have been exponents of reality from earliest ages. In each new epoch, they have propounded the enduring truths of practical living and have set out those standards by which men's lives may attain the highest achievements and fulfillment, generation after generation." *-Maharishi Mahesh Yogi*

Maharishi valued the wisdom of this ancient Vedic tradition of masters that is why he went to the Himalayas to find a master to serve. He found Guru Dev and served him nearly twelve years and continued his duty to him teaching Transcendental Meditation around the world for nearly fifty years until he passed away.

During the Ahshaya Triya celebrations in Vlodrop, Holland, in May 8, 2008, Tony Nader, MD, PhD, Maharaja Adhiraj Rajaraam, Maharishi's successor and first Ruler of the Global Country of World Peace, spoke of the ancient tradition of masters and called on his council of Rajas to serve. "We are the servant," he said. How inspiring to note that Maharishi's successor has embraced the same Vedic norm, service. Maharishi has followed that same norm in his lifetime in the service of Guru Dev, his master. Maharaja further said, "On this very auspicious day of Akshaya Tritiya, having received the blessings of the Vedic Pandits, and performed Puja (Vedic ceremony of gratitude) to Guru Dev, all our thoughts, feelings, and gratitude go to His Divine Holiness Maharishi Mahesh Yogi, who in the light of Guru Dev, in the light of the eternal Vedic tradition, has given everyone in the world the ability to achieve anything, to reach the highest goals and fulfill the highest aspirations of all sages and of all good-wishers of humanity."

"Maharishi has given answers to all questions, unfolded total pure knowledge, pure Being, in the lives of all individuals who learned his techniques and listened to his teaching."

"In life so many questions arise: Why are we here? Why are we living? What's the purpose of all of it? Why do we eat? Why do we have wishes and hopes and desires? Why do we see conflicts? Why and why and why—so many questions about life."

"And Maharishi has answered all of these questions. Of course, there are many answers on the individual life and different levels of the answers. Every individual has so many answers to so many of his questions, as there are so many individuals. And all of these answers are real on their level."

"The answers are also given in the Vedic Tradition, and the Veda, Itihas, Puran, Smriti; in Yoga, Karma Mimansa, Vedanta; and all the Vedas—Rk, Sama, Yajur, Atharva—all aspects of Shiksha, Kalp, Vyakaran, Nirukt, Chhand, Jyotish, they have answers—Sthapatya Veda, Ayur-Veda. All answers to different questions and different levels of questions."

"Yet Maharishi has given the one answer that the Vedic tradition upholds, that Guru Dev upheld, that Maharishi upholds, for the entire creation. This one answer is the secret of all secrets. That is the answer to every question. Ultimately every question that one asks about everything or anything comes to that one answer, to that one reality."

"And that secret of secrets is very simple. It is: *Sarvam khalwidam Brahm* -"Everything is Wholeness, everything is Totality, everything is pure Being." And therefore the quests of all quests, the search for all questions, the fulfillment for all desires, the achievement of all achievements, the highest aspiration of everything and everyone is to live the truth of life, the reality of existence. To fulfill that level of understanding that everything is Wholeness, everything is pure Being, pure silence, pure infinity, Totality, oneness. And that means myself, yourself, and everything that exists."

"Whatever we strive to do in life, whatever we want to achieve, is just in that direction, in the direction of reaching that understanding, and that level of experience in life."

Maharaja continued speaking on how the supreme understanding from the eternal Vedic Tradition of knowledge, the "great secret' that we have gained from Maharishi—that everything is *Brahm*, Totality—holds the answer to all questions in life. 'This knowledge is pure, simple knowledge, very simple knowledge—there is no more search anywhere."

"Of course it is not known by many people," Maharaja said. "Some people have heard it but they don't understand it, they don't believe it. Some people believe it but they don't understand it completely. Some people understand it and even believe it, but they don't know what to do with it!"

"Some people actually have had a glimpse of it, through transcending. Sometimes it has come to them in a moment of sudden enlightenment—just a moment of experience, be it a moment of joy with their family, with their loved ones, or with their devotion to God—and they had that beautiful moment, a glimpse of the wholeness of creation, of reality. And they don't know how to get it again, what to do with it? It was something, and—Maharishi gave this example—they may spend their whole life searching for it again."

"Some others, in doing good to others, in good actions, in feeling good—they have that inspiration, that sudden feeling of wellness, of unity, and they think, "This is the way to do it." That's one way to do it—there are many, many ways to achieve that. Again, the answers are on so many levels, every one of them being true and being real."

"Everybody is achieving steps in life, as if sometimes by trial and error, to do something and then learn—and then see when Nature has created a system, that if you do something in that direction you get more fulfillment, more happiness, and therefore you want to do it again; and if you don't, and it brings you other aspects of experience, then you don't want to do it, you want to go in (a different) direction—kind of trial and error. And learning from older people, more experienced people—learning from teachers."

"Then learning that time is also not a barrier, because some things can happen in a short time, immediate satisfaction. But there is also the long term, that Natural Law is also on a long term. You plant a seed today, and it might take years for the fruit to come and the tree to grow."

"But again these are ways of Nature teaching that as we go through time, and maybe many lifetimes—that reality is there as we approach it, and grow in it—as we get more fulfillment, more *Ananda*, more bliss. So (there are) different ways of experiencing it".

"Some know that reality in a more established way. They have transcended, they feel it, they experience it, as Maharishi has given the technology to transcend and directly Be That and experience That on a daily basis. Then in the growth of That, one comes to the realization, "I am That, I am that Reality." That's a great level of enlightenment. And beyond that, I start seeing others also as being That."

"So for myself, *Aham Brahmasmi* (I am Brahm, Totality); *Tat tvam asi*—"Thou art That," *Sarvam khalw idam Brahm*—"Everything is That." That is the highest level of realization, and that is the purpose of all doing, the purpose of all thinking, the purpose of all efforts to reach that reality."

"When living in the glory of that reality, as Maharishi has been with us, as the Holy Tradition is, as Guru Dev is, then comes yet one more level. What can be one more level than knowing oneself as Wholeness and everything is Wholeness? We have reached everything—what can be that one more level? That one more level is: making everyone else also realize that. Bringing it to the heart of everyone in the world, to the mind, to the intellect, to the understanding, and to the deep, true experience of that reality."

"That is the role of the Holy Tradition of Vedic Masters, which Maharishi has been for us, such a profound gift of Natural Law, which brought that understanding in its completeness—Total, unbounded perfect knowledge, perfect science of life."

Maharaja continued speaking on how, once having achieved the highest level of realization—knowing oneself and everything as *Brahm,* Totality—there is yet one more level: "bringing that realization to the heart of everyone in the world, through the gift of the perfect science of life that Maharishi has brought to us from the eternal Vedic Tradition of knowledge.

"And that science is not just pure theoretical knowledge, Maharaja said. 'It came from Maharishi with a technology—the technology of how to do it in a systematic and an orderly way. That's how we have seen this great enlightenment in the world, great experience of pure Being, of consciousness, of the Absolute which was missing in the lives of people."

"And, as we have seen that there are many who have not been able to understand, Maharishi has expanded his teaching to say "now you come into a group (of Yogic Flyers) and awaken collective consciousness." That way, even those people who are remote, who are distant, will feel that effect and awaken in that reality and then experience and want to search for it. And suddenly they will go to an introductory lecture on Transcendental Meditation and suddenly they will be interested—could be sometimes because they have high blood pressure, or whatever the reason, they are not sleeping well, something. Nature will bring them there. And then they will get into it as collective consciousness improves."

"He has also brought to us all the different aspects of Maharishi Sthapatya Veda, for example, that will help to organize structures of homes so people are more in tune with Natural Law, and therefore naturally will come to that desire, will understand more of the truth of life, and want to truly experience it and live it."

"Proper healthy food, proper healthy programs, avoiding things with side effects that disturb the functioning of the nervous system; and all of these different, different aspects - even Maharishi Gandharva Veda, beautiful music. All the aspects of proper food, agriculture, everything that is conducive for the physiology to align itself. And naturally as it gets more balanced, then the understanding and the clarity of thinking increase, and the individual desires the true direction, and starts seeing what the true purpose of life is—what the ultimate answer to every one of his questions is, and then the process of growth and evolution goes on."

"So where do we stand in terms of that reality of the gift of Maharishi today? In our perception, we see Wholeness everywhere, so therefore our role is based on the understanding and experience that everything is divine. Every human being is pure Being, is the light of God living on earth. Everyone has a physiology that is containing the Total Constitution of the Universe, so everyone is a temple of God, everyone is a living reality of Wholeness, of Totality, and therefore we adore everyone. It's a true adoration because we adore God, we adore infinity, pure Being, and everyone we know is that reality, is that pure Being."

"Now, in the functioning of the activities of our Global Country of World Peace, Maharishi has put us in that light of the Holy Tradition of Vedic Masters. Every time we were teaching transcending, Maharishi insisted it's in the light of the Vedic Tradition - we do Puja to Guru Dev."

"The technique is very simple - we all know, and those who don't know, they will get to know. But it's put in the light of a Tradition, because that is the connection that ensures that the teaching is connected to the source, and it takes the individual to the source. This is very, very important—the connection to the Vedic Tradition, the connection to Maharishi, to Guru Dev, to the whole Tradition of Vedic Masters—is very, very essential in terms of the long-term result of truly achieving the goal of individual enlightenment, and bringing that reality to the whole society."

"Therefore if we feel—Rajas, Ministers, (Teachers of Transcendental Meditation, and Yogic Flyers)—we are the representatives of the Holy Tradition, and we feel that everyone is absolute pure Being, and therefore we have adoration for everyone; and it's only from that platform that we take our action as guided by the Holy Tradition, by Maharishi, to bring to the world that perfect knowledge of life."

Maharaja Adhiraj Raja Raam commented, regarding the role of the Rajas, that it's not a question of Rajas wearing crowns or speaking and acting according to ceremonial procedures. 'We are here to adore and to serve. That's our role, *we are the servants. We are the servants."*

"But in every kind of situation one has many hats. The man who goes to his job, he has children. At home he is the father. When he drives the children to home, to school - he is the driver. When he is the judge he goes to the court, he's the judge - he's a very important person. So he performs these roles in life. Yet, if he is enlightened, he knows that whether driver or judge - it doesn't matter - he is pure Being, Absolute pure Being, like everyone else is."

"And that's how we feel. So we do have a system, we do have etiquette if you like, because the judge has to give judgment in the courtroom, and therefore he has to have certain dress, a certain way of doing things. When he's driving he has to be careful, has to follow the rules of the road, to stop on the red light. There are rules he has learned that he has to follow."

"So there are systems that ensure that there is a proper relationship that allows knowledge and experience to flow in the proper way. It's never on the level of who you think you are, who you think I am. What I am, what you are, who is where? Humble, more humble, honest, simple, innocent—all of these are real on many levels, but on the level of truth of life, it's only that one reality. One unbounded pure Being is the reality of everything."

Maharaja continued discussing the organizations and systems established by the Global Country of World Peace that ensure that there is a proper relationship that allows knowledge and experience to flow in the proper way.

"Then there is a system of flow of information, of structure. That's why Maharishi has created a system such as our system - our system, in the light of the Holy Tradition of Vedic Masters, where everyone does carry a responsibility, everyone has a duty on that surface level. It's a surface level, that's why they call it like a *Lila* (play) - it's like a theatre. It's not that we are acting in the theatre . . .

but you imagine someone going into the theatre, and then suddenly he is to play a certain role, and that role he plays fully, and integrates it, and he believes in it."

"So in the *Lila* of life, every one of us is playing a role, every one of us is putting himself in a position of a role to play. Of course, the actors are also selected according to how they have certain abilities to act in some way or another. Some may have more experience in architecture, some more in health, some more in finance, some more in management, some more in administration. And it's the same one pure Being playing these different roles."

"So the Minister of Health doesn't really feel in his deep Self, he is different from the Minister of Architecture, because he knows it's one and the same. Yet they are playing different roles. That same person, that same individual is playing these different roles. So we are here to give enlightenment."

Maharaja Adhiraj Raja Raam said. "We are here to fulfill Maharishi's most supreme achievement of all times, for having brought total science and total technology to create true enlightenment in the world, and truly achieve the lasting—everlasting achievement, which is alone Pure Being; because there is nothing that lasts except pure Being and its expressions. Its expressions, when they are in tune with it pure Being—when the house is in tune with Natural Law, it is more lasting and brings more fulfillment. When the individual is purely, even on the physical level, truly fully integrated, the Constitution of the Universe is integrated in his physiology, therefore he can live longer, he can last longer. Even immortality has been mentioned in the Vedic Tradition. That is possible according to time, and according to requirements of Natural Law and the quality of time."

"Therefore I wanted today, in explaining all of this, to give all of you and the whole world a pledge from my side: the pledge that every one of my intentions, feelings, thoughts, planning, be in that every simple guiding light of Maharishi to create Heaven on Earth."

"It's a beautiful feeling . . . . I do feel that any minute we wait in the ability to teach someone Transcendental Meditation and bring to him enlightenment, even some glimpse of the Absolute, is a minute wasted if we don't use it for that."

Maharaja explained that he would be meeting on that day of Akshaya Tritiya, the Day of Lasting Achievements, with the Prime Minister and the Rajas and Ministers of the Global Country of World Peace, because we feel that urgency today to bring to the whole world that light of knowledge."

"We don't want our world to go into any pains and difficulties. We are aware that there are problems, we're aware and feel saddened by natural disasters in Myanmar, and the loss that is going on there; and any wars, also any conflicts in the world. We know that we can create Heaven on Earth; therefore I personally call—and I know all my Council and Ministers and Rajas with me - call upon everyone in the field, everyone who practices Transcendental Meditation, every Yogic Flyer, every Teacher of Transcendental Meditation, to come together and think of the highest achievement that one can do today is a lasting achievement."

"Think that all these details of anything are of no importance. Today, the world needs you. You have been put on that level, the highest level of possible human consideration and dignity. Not only that level where you know even intellectually, and a lot on the level of experience, that you are *Brahm*, you are Totality and everyone else is that Reality. You have been put on an even higher level. You have been put on the level that you can give that knowledge to others. That is even higher. That is the level of the Guru, the teacher, the giver of reality, the Holy Tradition of Vedic Masters."

"So when a Teacher of Transcendental Meditation has received that gift, then it's a big responsibility. No other consideration of past, present, or anything, should come to our mind, except to come together, go out, and bring enlightenment. That is how we can fulfill Maharishi's desire to create Heaven on Earth, Raam Raj (Administration through Natural Law) for all mankind."[43] Maharaja Adhiraj Rajaraam adhere to service, it will guarantee Maharishi's achievements to last for many generations to come.

---

[43] *May 8, 2008, Celebrating Akshaya Tritiya, day of Lasting Achievements*

*His Divinity, Brahmananda Saraswati,
Maharishi's Master, Guru Dev*

THE HOLY TRADITION OF MASTERS

18

GLOBAL CLIMATE TRANSFORMATION

It was Christmas time, in 2008, some years ago. I was driving to the Los Angeles airport to pick up my daughter, who was coming to California for a visit. It rained nearly a week. All the roads were drenched with so much water and traffic was nearly a standstill all the way to the airport. The grey clouds hovered in the sky making the afternoon stark dark even before sundown. The sun seemed to have disappeared nearly a week. The early afternoon sky looked so gray with heavy clouds hovering above.

The plane was delayed an hour. From the airport, on the way to our destination somewhere near the Magic Mountain area in Valencia, California, my daughter was bewildered by the rain. She said, "Mommy, never in the past fifteen years that we have been coming to visit you on Christmas, there was never this much rain."

I answered, "The recession here has caused many problems. Home foreclosures and unemployment is a problem that disturbs many people's mind that might have caused these climate disturbances and imbalances in nature."

Perplexed, she said, "You never stopped correlating natural disasters to the disturbances in the human mind." I was hesitant to narrate and correlate the behavior of finest particle in the human mind and the particles in the atmosphere to my daughter. It is a long complex explanation and I did not wish to complicate her excitement spending the holidays with me. Her husband Paul and her two kids enjoy being in California every Christmas season instead of the freezing December temperatures in Canada where they live. I did not wish to describe and present to her the long illustration of how the mind works, like how messages sent from the International Space Station reaching the earth work through a field or the millions of emails that pass thru the atmosphere around the planet in seconds thru a particle field that is interconnected with each other and how thoughts relate with the finest particles of nature found within the human mind.

Instead I replied, "I am only repeating what Maharishi was saying about quieting the mind and nature will stay quiet too. Nature has roots within the human mind at the level of the Unified Field. When a group equal to one percent of the population of the country practice basic Transcendental Meditation or the smaller number of the square root of one percent population practice TM and the advance TM-Sidhi program in one place together, climate balances and disasters are averted."

"Imagine only about 2,500 group here in the US can balance the climate. It is just a small group compared to the tremendous amount of money lost in floods and many lives can be saved. Deficits, drought, hurricanes, and all the negativities happening here is depressing. It is even more depressing knowing that a solution is available, so few seem to care about Maharishi's unconventional remedies to solve these calamities. All the previous experiments showed balancing of the climate, not weeks of heavy rain. There are so many people suffering this Christmas time. There are homes flooded and eroded down the mountains," I replied sadly.

My daughter said, "Do not be sad, Mommy," my daughter comforted me. "There is nothing you can do." I said, "I wish somebody will listen to Maharishi's call and use his technology as a disaster control and prevention."

"Few listen because it is hard to believe that the mind has the ability to control natural disasters," she commented. "You are the only one I know who believes this Maharishi thing, controlling natural disasters."

I narrated to her, "when I was building the thirty-storey tower in Manila, I tried this Maharishi thing several times. It appeared true from my point of view, having experienced climate changing each time I asked the four hundred workers to meditate. Many times when we were pouring concrete and there is rain, I would call all the four hundred workers to stop work and meditate. The rain stopped. It happened several times. The workers were laughing when the rain stopped each time. At least the work continued, otherwise the work will stop as the rain will dilute the concrete mix and we cannot pour the concrete if rain water dilutes it. More water will lessen the strength of the concrete. We need to have test cylinders for every load."

"We cannot stop pouring concrete that was previously ordered either, as the concrete will harden in trucks down below and I have to pay for it just the same. So I had no option but to make the workers meditate to stop the rain, it did stop each time. The workers liked it. Otherwise work will stop and they will not get paid if they do not work. The construction workers were paid on a day to day basis. I had no other solution but to use TM. So, I believe what Maharishi has said, climates balance with large groups quieting their minds. TM quiets the mind. It happened many times. How can I not believe it."

The workers laughed every time the rain stopped, but it did stop and the worked continued." I recounted the events during the construction of the building.

"The first time it happened, I was skeptical too, like everyone was. We repeatedly did it, and the result was repeatedly the same every time. I felt really awed by these occurrences. The occurrences made by belief very strong and nobody among the workers can refute the outstanding results that everybody saw happen. Even in the schools I have previously built manifested the same climate change. How can I not be affected by these negativities occurring? I know of a solution, but now, it makes me really sad, seeing all the rain and listening to all the floods going on, unable to do anything about it, I have grown old."

"I have limited resources now. My previous experiences strengthened my belief in Maharishi's technology. It worked all the time, not some of the times. It happened also when all the students meditated when the temperature was very warm, it rained and balanced the warm climate, it happened not only once. Every time I opened a new school when the temperature was high in July, it rained. In large groups together, it activates the temperature mechanism built within the human mind, when warm, it cools the environment, when cold, it warms it. It is the law of physics in thermodynamics."

"Maharishi's theory was right, TM balance the climates. I was a witness to it several times, having repeatedly done it with the four hundred workers and with the schools I have built, so I know it is true. I believe it. That is why, it pains me every time natural disasters like this happen and there is nothing I can do. It is difficult to convince people, not only ordinary people but especially people with the capability to form a group. It should be a governmental effort," I said. "Because the government is in charge of rescuing people in flooded areas and rebuilding destroyed structures. A lot of effort and money lost. It is mostly governmental effort to rebuild roads and infrastructures destroyed by floods."

"There are few individuals who take the responsibility to establish a group. The Vedic pundits of India is being undertaken by donations from many groups, small donors and big ones. There was one person with the means here in the US, Dr. Howard Settle in Fairfield, Iowa. He supported TM practitioners to meditate as a group, but so few stay for long. People come and go because they need to go home and work. A permanent group needs to be in place."

"At one time there was a strong hurricane coming and it reversed its course soon as groups started to fly at the Domes located at the campus of the Maharishi University of Management in Fairfield, Iowa. I heard it happened sometime in 1986."

"There are just a handful, few groups who can stay on a permanent basis. The problem is on both sides, both sponsors and people. There are only a handful of sponsors and fewer meditators who can stay for long. The need for sponsors and meditators working together surmount the challenge that will benefit the whole population."

"I think it should be a governmental effort as there are thousands of soldiers and police force already under payroll in every country. One government has took it one time, President Chisano of Mozambique in Africa, in 1994. Rain fell in the drought areas as soon as large groups in his military started to meditate. With all the problems around the world going on, you cannot coerce people to believe. It is like forcing a horse to drink, if it doesn't want to drink."

"There is now in India a group of more than 1,000 referred to as Maharishi's Vedic pundits at the central of India," I related the status of the India project to my daughter.

"It took Maharishi more than fifty years speaking of the knowledge. Now that he is gone, who knows if the knowledge will continue?" my daughter argued. "There will always be these natural disasters happening until a group will be formed to balance it," I said.

"Try the Maharishi Effect and see what happens with earthquakes, fires, typhoons, and wars. It can avert so much loss and destruction. The cost to do it is minimal compared to the incomparable damages they produce."

It was New Year's Day, 2011. In one day, thirty six tornadoes created havoc and devastated six states near the eastern part of the United States. There were many homes destroyed on New Year. I listened to the news speechless. I could not say a word as people around me have yet to believe Maharishi's system of control and prevention of natural disasters. The ancient Greeks scholars called storms and tornadoes, "particle turbulence," which is actually strong air pressures caused by cooling off air particles and turning to rain. Their rapid movements cause by the air imbalances, the cooling of warm air by the CO_2 layer in the atmosphere rapidly absorbing the warm air from the sea cause the strong winds that destroy many structures. The thicker concentration of CO_2 the greater destruction is created.

At the same time, in Australia, the worst flooding in 50 years raged in biblical proportions in Queensland State and adjacent cities, in an area as wide as France and Germany combined. In Brazil, rains, landslides, and floods devastated homes never experienced in 25 years. It was a season that was supposed to be a time of happiness, as it was Christmas time and the New Year of 2011 had just begun.

A few months later, on March 11, 2011, an earthquake rocked Japan. The devastation was unthinkable! The tsunami swallowed and swept away everything along its path, homes, cars, and a whole town washed away into the sea. The toll in lives and properties destroyed was enormous. I felt devastated in spirit as the world watched in awe. The destruction of the nuclear power facilities posed so much more harm than the earthquake itself.

The hopelessness of the moment, seeing so much destruction, engulf my heart with sadness and I just wished Maharishi's knowledge will come into use one day soon and vanish away all these sufferings. People work and live to expand fortunes and material possessions. But who knows, in one single stroke of a natural disaster, like this one in Japan, ones possessions were swept away into the sea.

Until such time, scientists will continue to research and publicly publish the effects of the Unified Field occurring within most silent state of the human mind thru the practice of Transcendental Meditation and Yogic flying in large groups together.

I hope that governments will pick up the knowledge and implement it on permanent basis, then order in nature will follow the order and harmony of the functioning of the universe also found within the quiet level of the human mind. Then, it will be everyone's endeavor to work for Maharishi's greatest knowledge to protect every ones family, fortune and future. Transcendental Meditation is so simple and easy as taught by teachers trained under the guidance of Maharishi.

It has shown awesome results described and validated by many scientists, in contrast to other types of meditation that makes the mind active and stay on the surface level unlike other concentration or contemplation techniques. TM quiets the mind and does not concentrate on the heart beating or the lungs breathing cycles. It is a unique procedure of quieting the mind. So, no matter how many different churches are filled with people or number of times a day they pray, it doesn't matter, the difference with TM and other techniques is that it quiets the mind in a simple easy way.

It always works and you do not need to invoke Maharishi's name or the name of any god. TM does not refer to any religious doctrine or belief. It is a natural law phenomenon. Balance and order in nature follow the same order occurring when the human mind quiets down and draws the Unified Field which is the seed form of natural law.

The solo particles of nature, the primeval atom, is more powerful than the hydrogen atom, it is invisible and indivisible which permeates the human mind, like the sap inside the tree that permeates from its roots to the branches and its leaves, they permeate the atmosphere and create the order, peace and prosperity for all.

There are many groups who wish for peace and pray for peace since time immemorial. Many centuries have passed and still the condition of our planet has not changed and wars and wild typhoons still occur. "The Maharishi Effect needs to be in place for disasters, calamities and crimes to be controlled and prevented on a permanent basis."

I continued driving carefully in the rain, it was past midnight. My daughter was reading the street names to me so that I will not get lost driving in the rain at night. Then I said and joked, "If somebody says there are diamonds enough to feed the world in the mountain at the end of the road, many will try to find it. A new gold rush will happen again in California. Maharishi's discovery of the Unified Field, is more than a diamond discovered. It can remove floods, droughts and fires. It can solve global warming. It will create peace and prosperity for every country. How amazing. A rush to implement it should be done."

"Maharishi said that after the nine thousand is formed, the world will prosper and be at peace, only a handful try to form a large group. For now, I know of only two large groups, in Fairfield, Iowa, USA and at the central part of India. A few small groups pop up here and there."

The reaction of people depends on the material benefit they get, not what benefit others get from it, the Unified Field give to both the benefits. In the gun control issue, there are so many shootings and killings happening around, in schools, in theaters, in public places and even in house of worship. Those who profit from it are against gun control. The person holding the gun is the problem. If he is stressed and angry, he can shoot. TM releases anger and negative emotions. TM makes one happy. The target of recourse for gun control should be the holder of the gun, not the gun. Gun control should be in controlling of the anger of the individual. TM removes anger. No happy person can shoot another.

Many celebrities and people in the media are practicing TM and they benefit from it. I wonder, few of them should lobby governments to utilize it. Maybe because the Maharishi Effect has to be experienced many times before one can believe in it. Like I did a couple of times, otherwise I will be hesitant to talk about it too."

I experienced the Maharishi Effect several times, that is why I truly believe in it without any hesitation. Even if many research have been published around the world, many still are skeptical because you have to experience and monitor the effects for yourself to believe in the Maharishi Effect and its positive results. Maybe soon, when governments find no solution to the magnitude of problems, they will try this one unconventional system of balancing of climates.

The Philippines is a typhoon zone and at least twenty typhoons happening and growing stronger each year. I remember when there were students and workers from my team who were meditating together I noticed there were less typhoons occurring. The super typhoon Haiyan should be a wake - up call for the governments. They should explore all options to prevent another of its kind happening.

During my time, we spoke to then President Marcos about the Unified Field. I spoke about it to the generals in the army. We got an instruction from Maharishi to see President Marcos before his fall. I was able to obtain a direct line to the troubled President from a leading newspaper editor who had learned Transcendental Meditation. I told him Maharishi wants us to see Marcos, the President. "The country is turbulent and needs a group. It is like a house that is burning, go," Maharishi instructed us.

The country was in turmoil. In November 1983, the prominent opposition leader, Benigno Aguino, was assassinated at the Manila International Airport as he arrived from a trip from the US. The assassination caused rallies and protests in the streets. Maharishi was concerned seeing all of nature's signals, an earthquake, a volcano and a wild typhoon occurring simultaneously in the same period. Nature was giving signals that needed action and recourse. Maharishi foresaw the turbulence of the people's mind, the protests that followed are nature's signals. Groups of people gathered in the streets yelling, "Marcos, step down!" I called for an appointment to the secretary at the Malacanang Palace where the President stayed then. Marcos asked his generals to listen to us. The generals laughed at us instead. They interrogated us with sarcasm as we delivered Maharishi's message that the turbulence in the country would settle down if there was a group meditating together.

"Like boiling water, lower the temperature and the boiling water will instantly simmer down. Lower the turbulence in the mind of a group through Transcendental Meditation, the people protests that is happening now will simmer down," I explained. Maharishi knew what action to take.

When Maharishi sent the twelve hundred TM and Yogic flying groups to the Philippines, he must have known that the turbulence in the human mind created all these nature related problems. A group equal to 1,200 TM and Yogic flying participants were needed to quiet down the turbulence in the environment.

Einstein once mentioned that the mind creates the environment. Both physicists, Einstein and Maharishi knew of the effect of turbulent minds. Maharishi knew the solution, that is to quiet the minds of a large group of people practicing the TM and Yogic flying program equal to the square root of one percent of the population of the Philippines. That is why Maharishi sent one thousand two hundred meditators to the Philippines at that time of history of the country to quiet their mind with Transcendental Meditation. There became a peaceful transition to a new government.

Maharishi said, "The human mind is the seat of nature." The seed of nature is enlivened in the human at the level of the Unified Field experienced during the TM practice. Not a single one individual was killed during the government transition to a new ruling party. With all the wild riots that happened there was not a single casualty. This was in contrast to many fatalities in countries like Libya, Egypt, Ukraine and in Syria thousands of people die in the attempt to overthrow the existing Assad government.

Somewhere around the world some war will pop up when people are angry and where human rights are violated. Some militants will suddenly rise. Mobs formed. There are protests here and there in some troubled parts of the world, more so now than ever.

People protests are frequent occurrences and must be prevented. They can bring down the fall of leaders. It did with former President Marcos of the Philippines at that period of time in the Philippines history.

It caused the fall of Hosni Maburak of Egypt and Kaddafi in Libya and Yanukovich in Ukraine. Protests are signs of citizen discontent. It is a sign from nature too, that the minds of the people are turbulent and needs to quiet down.

Man's birth right is to have peace. Silence is within the reach of the human mind. The mind is the instrument of peace. Live in accord with nature with Transcendental Meditation and benefit from the technology of the Unified Field and its spectacular group effect.

In a scientific experiment, introduce a new element when nothing appears to be working. Nothing has worked for peace in modern times, even in the past, nowhere can you find one system without the use of expensive military arms. TM can relieve anger, frustrations and negativities. Like water lowering its temperature at zero degrees, impurities moves out of the ice crystals. Man's mind at zero activity, negativities disappear. The military of each country is the perfect tool for the Maharishi Effect because soldiers are already available and their duty is to defend the country.

The Maharishi Effect will provide the invincibility of every country on a permanent basis. No enemy will be born. No destruction will occur. The Maharishi Effect and the military group implemented will provide the peace and economic prosperity for every country.

When man reaches the state of unity consciousness on a permanent basis, all his thoughts and actions will become in accord with natural law. TM paves the way for the evolution of man in the direction of all good.

19

MAHARISHI'S ACHIEVEMENTS

Maharishi's achievement is unparalleled in history. Maharishi's discovery of the Unified Field brought to light the unseen, non-material aspect of the material form of creation, the very basis and root of all matter forms in the universe.

The basis of the future of scientific endeavors will have a significant impact based on the discovery of the Unified Field, the supreme and ultimate science of man and his surroundings. With the application of the Unified Field mankind will enjoy life, a heaven on earth as Maharishi has predicted.

The human mind has the capability to balance climate as Maharishi has foretold. Before Maharishi brought out the Natural Law Principle of Invincibility,[44] no one knows how to solve natural disasters like floods, typhoons and volcano eruptions constantly occurring on earth. No one knows the answer. The only answer we hear often when natural disaster occur, most say, "That it is will of God." Maharishi has identified the Principle of Invincibility which can be utilized to balance climate changes and reduce massive distructions brought about by typhoons, droughts and even earthquakes and volcano eruptions.

The Philippines is a typhoon prone country, nearly two dozen of them occur every year. The government should consider Maharishi's Natural Law Principle of Invincibility to avert the many typhoons constantly occurring in the country causing huge destruction and death among its citizens.

The United States, a lot of fires, hurricanes and tornadoes occurring. It has the means to review and consider Maharishi's options and refer to the knowledge validated and preserved in Fairfield, Iowa, USA at the Maharishi University of Management.

The invincible Unified Field effect that Maharishi has brought to light can solve climate turbulence. Unconventional as the solution may seem, implementing Maharishi's principle will prevent and save billions of properties destroyed by floods and typhoons.

Maharishi's achievement is an easy and simple solution to climate change that is an unprecedented discovery in human history.

---

[44] *Natural Law Principle of Invincibility, See Chapter 14*

Not anymore the name of God is called to solve problems of climate changes, but call on Maharishi's Technology of the Unified Field. Man has the intrinsic capability to balance climates with temperature balancing mechanism built in the human mind at the level of silence in large groups together. With the onslaught of the discovery of the Unified Field, God really has provided all the tools man needed for fulfillment beyond suffering that man himself holds. If the ancient Vedic civilization was the first origin of man, God provided the sounds of natural law contained in the Veda and Vedic literatures prevailing in the human physiology that can avert any destruction brought by wars and climate changes.

These sounds of natural law, the Vedic sounds were chanted and passed on from the ancient Vedic families generation to generation. Veda is the self or the self is the Veda. The sounds of the Veda describes the same frequency located within the human physiology that spring from the non-material form that emerge to become the DNA, then to the material form of cells, blood and matter forms into muscles and bones, ith the same frequency emerging to become matter forms seen in all of creation.

Transcendental Meditation utilizes meaningless sounds called mantras. They are sounds of natural law which balance the human physiology as well as balance the natural rhythms of the surrounding environment during the practice of Transcendental Meditation and the Yogic Flying in large groups together.

The sounds of natural law governing the whole universe also governs the human mind and body. The human mind and body has its basis in the sounds of natural law contained in the Veda and Vedic literatures.

Medical machines like the modern ultrasound machines measure the frequencies of human body including the EKG and EEG machines that measures brain frequency and heart rhythms. They illustrate that indeed sounds does prevail within the human physiology.

The sounds of natural law are the basis of all life. Without the rhythms of the human breathe, life ends. The Veda, that contains the rhythm of life also contains the total knowledge man needed to use at its disposal to live life that is problem free. But due to long passage of time, the knowledge was lost and forgotten. Maharishi brought it back. He gathered all the lost and fragmented knowledge of the Veda and Vedic literatures and made it into a complete Vedic science.

Maharishi called on modern scientists to validate the ancient

Vedic sciences thru the discovery of the Unified Field. Now physics has unfolded the scientific meanings of the Veda and the Vedic literatures thru the discovery of the Unified Field. The ancient Vedic sounds became significant in the improvement of health in the application of Maharishi's Vedic Vibration Technology that can provide instant relief in some chronic diseases that modern medicine find it difficult to completely eliminate such as high blood pressure, hypertension, diabetes and migraines.

Maharishi's achievements [45] is truly significant in bringing out the supreme wisdom contained in the Veda and Vedic literatures which spanned nearly fifty years since Maharishi came out of the Himalayas in 1955 until his passing away in February 5, 2008.

Maharishi endeavored to restore the Vedic traditions to the modern world. These traditions mark the character of man as a cosmic being enjoying heaven on earth for all times without problems and suffering.

Maharishi's goal when he came out of the Himalayas was to restore the ancient Vedic norms of life and duplicate it to the whole world. These norms were too significant to be ignored as it tremendously transfoems the well-being of mankind.

Maharishi gathered the fragmented Vedic texts that contained the ancient wisdom to imrprove man and his environent into a complete science of consciousness, the Maharishi's Vedic science. The ancient Vedic texts spoke of man's birthright to happiness and fulfillment beyond suffering. Maharishi wanted to bring the destiny of man to its highest status to complete fullness in evolution gaining happiness and prosperity.

Maharishi's commentary of the Bagavad-Gita, he said, " The Veda reveals the unchanging Unity of life wich underlies the evident multiplicity of creation, for Reality is both manifest and unmanifest, and that alone is, "I am That, thou at That, and all this is That' is the Truth; and that is the kernel of Vedic teaching which the Rishis extold as teaching worthy of hearing, contemplating and realizing. The purpose of this commentary is to restore the fundamental truths of the Bagavat-Gita and thus restore the significance of its teachings. If this teaching is followed, effectiveness in life will be achieved, men will be fulfilled on all levels, and the historical needs of the age will be fulfilled also."[46]

---

[45] *maharishi-programmes.globalgoodnews.com*

[46] *Maharishi, 12 January 1965*

In his own words, Maharishi described Vedic Science, "Vedic Science is the science of Veda.' Veda' means pure knowledge and the infinite organizing power that is inherent in the structure of pure knowledge. Pure knowledge is the state of awareness in which consciousness knows itself alone, when awareness is completely self-referral, when awareness has nothing other than itself in its structure. This state of pure knowledge, when knower, known, and process of knowing are in the self-referral state, is that all-powerful, immortal, infinite dynamism at the unmanifest basis of creation."

"This is Vedic Science: all about the knower, the known, and the process of knowing; all about consciousness, both in its self-referral, self-interacting state, and as it expresses itself in the infinite variety of the whole creation— that performance of Nature which goes on and on eternally in all spheres of time—past, present, and future. The knowledge of this most basic principle of life—the self-interacting dynamics of consciousness is the science of pure knowledge, the Science of Veda."

"Human awareness can identify itself with this most basic, self-referral value of consciousness in the state of Samadhi, or Transcendental Consciousness. This state is easily gained and most naturally enjoyed through Transcendental Meditation. The functioning of transcendental pure consciousness is the functioning of Natural Law in its most settled state."

"The conscious human mind, identifying itself with this level of Nature's functioning, gains the ability to perform in the style with which Nature performs its activity at its most fundamental level. Completely identified in Transcendental Consciousness with the full potential of Natural Law, the human mind is a field of all possibilities, spontaneously functioning in harmony with all the Laws of Nature and able to accomplish anything. This is how human life will be cultured through Vedic Science."

"What is the difference between Vedic Science and modern science? Through its objective approach, modern science reveals that which is perceived, the object. The subject, the perceiver, remains separate from it. Modern science investigates the field of the known, but it does not touch at all the field of the knower and the spontaneous process of knowing."

"Vedic Science is a complete science, which extends and fulfils the objective approach of modern science by incorporating the knower and the process of knowing into the field of investigation. It provides a complete and comprehensive knowledge of the Unified Field of all the Laws of Nature, which can be described as the unified state of the knower, the known, and the process of knowing. Vedic Science also describes the sequential mechanics through which this three-in-one structure of the Unified Field gives rise to the infinite range and diversity of Natural Law displayed in the universe."

"The knowledge of the Unified Field has been discovered by modern science during just the last few years, but complete knowledge of the Unified Field has always been available in the Vedic Literature. Today Quantum Physics has glimpsed the details of the Unified Field and is locating its three-in-one structure. This is precisely the three-in-one structure of the self-referral state of consciousness."

"Credit must be given to modern science because its objective approach has now uncovered the reality of pure subjectivity. The world of scientists should know, however, that the objective approach comes to an end there. Many more discoveries may be made on the surface, relative levels of existence, but the goal of Physics has been reached."

"Today the most advanced level of modern science needs a complete approach to investigation, which includes the two other values of knowledge—the value of the knower and the value of the process of knowing. All three values are uncovered in their totality in Vedic Science."

"If progress is to continue, a shift is required from the science of only one category to a total science. Vedic Science is that total science. It uncovers the knowledge of the total potential of Natural Law in its completeness and brings human awareness in tune with those fine creative impulses that are engaged in transforming the field of intelligence into the field of matter.

"The approach of Vedic Science is such that the knowledge gained enlivens that most fundamental value of consciousness from where all thoughts and actions emerge. Therefore, the very methodology of gaining knowledge through Vedic Science is such that as one gains the knowledge of Natural Law on the intellectual level, one begins to live that Natural Law in daily life in a most spontaneous way. This is the basis of the practical application of Vedic Science."

"Vedic Science is applied through the Technology of the Unified Field. We speak of the Unified Field in connection with Vedic Science because of the similarity between what has been discovered by Physics and what exists in the self-referral state of human consciousness. The Technology of the Unified Field is a purely scientific procedure for the total development of the human psyche, the total development of the human race. This is the time when objective, science-based progress in the world is being enriched by the possibility of total development of human life on earth, and that is the reason why we anticipate the creation of a Unified Field-Based Ideal Civilization."[47] —*Maharishi*

Maharishi on Transcendental Meditation

"My Transcendental Meditation is a simple, natural, effortless procedure whereby the mind easily and naturally arrives at the source of thought, the settled state of mind—Transcendental Consciousness—pure consciousness, self-referral consciousness, which is the source of all creative processes."

"This process can be likened to a river which naturally and effortlessly flows onto the ocean and gains the status of the ocean. 'Transcendental Meditation is practiced for 15 to 20 minutes in the morning and evening, while sitting comfortably with the eyes closed. During this technique the individual's awareness settles down and experiences a unique state of restful alertness; as the body becomes deeply relaxed, the mind transcends all mental activity to experience the simplest form of awareness—Transcendental Consciousness—where consciousness is open to itself. This is the self-referral state of consciousness."

"The experience of Transcendental Consciousness develops the individual's latent creative potential, while dissolving accumulated stress and fatigue through the deep rest gained during the practice. This experience enlivens within one's awareness creativity, dynamism, orderliness, and organizing power, which results in increasing effectiveness and success in daily life."

"Transcendental Meditation can be easily learnt by anyone. People of all levels of intelligence, belonging to all ages, belonging to all cultures, religions, and educational backgrounds in countries throughout the world practice the technique."

Since the beginning of the Maharishi's Worldwide Transcendental Meditation Movement in 1957, over six million people have learnt Transcendental Meditation and forty thousand teachers of Transcendental Meditation have been trained throughout the world by Maharishi.

"The Transcendental Meditation Program is the most direct way for the fulfillment of the supreme philosophy of life, where the individual lives life in full enlightenment, and where the individual breathes life in Cosmic Reality."

---

[47] *Maharishi's public lecture on Vedic Science*

"As the practical aspect of the supreme philosophy of life, my Transcendental Meditation is most natural, and as such it is universally applicable for the well-being of the individual and society as a whole."

"Transcendental Meditation renders all aspects of life according to Natural Law. Its application in the fields of politics, economy, religion, and culture of every country is so rewarding that it has started to bring a new wave of excellence to lead the present civilization to the height of perfection."

"The benefits of Transcendental Meditation are so many and so great that it has been our delight to launch a program for the complete transformation of life on earth—Heaven on Earth. How is Transcendental Meditation able to accomplish this overall enrichment of life?"

"Firstly, Transcendental Meditation is the Technology of Consciousness, which is the most basic element of life - the home of all the Laws of Nature. Transcendental Meditation unfolds the creative genius of the self-referral state of intelligence—the home of all the Laws of Nature—and inspires the Laws of Nature to uphold all thought, speech, and action: support of Natural Law increases in daily life with regular practice."

"Secondly, it eliminates the most basic cause of all stress and strain. Everyone lives his life through the cycles of routine work, whether he is a student, a working adult, or retired—the whole population is living life through routine work, and routine work in daily life does not provide an opportunity for the full expression of Creative Intelligence. This lack of opportunity to display creativity causes frustration and becomes the basis of all anti-social behaviors"

"Transcendental Meditation helps the awareness to transcend boundaries and go beyond the field of limitations; this is how Transcendental Meditation, providing the opportunity to create unbounded infinity, eternity, satisfies the inner creativity and inner genius of life by providing an opportunity for the full expression of Creative Intelligence. The daily opportunity for the individual's awareness to go beyond boundaries (through Transcendental Meditation) neutralizes the rigidity caused by the boundaries of the daily routine."

"Thirdly, the marvel of Transcendental Meditation is that both of these above-mentioned values—the blossoming of creativity, and dissolution of stress and strain—are achieved in one stroke."

--Maharishi

The TM-Sidhi Programme Including Yogic Flying

"The TM-Sidhi Programme is an advanced aspect of Transcendental Meditation. It trains the individual to think and act from the level of Transcendental Consciousness, greatly enhancing the coordination between mind and body, and developing the ability to enliven Natural Law to support all avenues of life to fulfill one's desires."

Yogic Flying

"It needed a scientific age for the world to appreciate the significance of the philosophy of Yoga and its practical application in creating integrated individuals, integrated nations, and an integrated world family."

"Yoga means union, the union of the individual awareness with the unified field of all the laws of nature in the state of transcendental consciousness. Yogic Flying demonstrates the ability of the individual to enliven the total potential of natural law in all its expressions — mind, body, behavior, and environment." 'Yogic Flying' presents in miniature the flight of galaxies in space, all unified in perfect order by Natural Law."

"The mind-body coordination displayed by Yogic Flying shows that consciousness and its expression — the physiology— are in perfect balance. Scientific research has found that maximum coherence in human brain functioning gives rise to Yogic Flying. As the coherently functioning human brain is the unit of world peace, Yogic Flying is the mechanics to make world peace a reality, and thereby bring world health, world happiness, world prosperity, a world free from suffering — Heaven on Earth in this generation."

"Yogic Flying is a phenomenon created by a specific thought projected from Transcendental Consciousness, the Unified Field of Natural Law, the field of all possibilities. This is the simplest state of human consciousness, self-referral consciousness, which is easily accessible to anyone through Transcendental Meditation, and is enlivened through the TM-Sidhi Program, which leads to Yogic Flying. Yogic Flying demonstrates perfect mind-body coordination and is correlated with maximum brain wave coherence (EEG), indicating maximum orderliness and integration of brain functioning. Even in the first stage of Yogic Flying, where the body lifts up in a series of short hops, this practice gives the experience of bubbling bliss for the individual, and generates coherence, positivity, and harmony for the environment."

"Regular practice of Yogic Flying leads the individual mind to enjoy control of Nature's central switchboard, from where Natural Law governs the life of everyone and administers the entire universe from within the intelligence of every grain of creation. From this level of the total potential of Natural Law the individual can command all channels of Nature's infinite creativity and the invincible organizing power of Natural Law. Just as an order from the Prime Minister commands the total authority and resources of the nation for its implementation, similarly, any intention projected from the Unified Field of Natural Law commands the infinite organizing power of Natural Law for its immediate fulfillment."

"The practice of Yogic Flying provides a practical demonstration of the ability to project thought from the Unified Field of Natural Law, and develops the ability to act spontaneously in accord with Natural Law for the fulfillment of any desire. The phenomenon of Yogic Flying proves that through my Transcendental Meditation and TM-Sidhi Program, anyone can gain the ability to function from the simplest form of his own awareness and can develop mastery over Natural Law."

"My TM-Sidhi Program provides a direct entry to the full blossoming of the creative genius of everyone; it is a master key to open the field of higher states of consciousness, where one naturally lives life supported by the evolutionary power of Natural Law.

"All parents, all teachers, and all governments should make these programs available to the growing youth in order to save them from problems and suffering in life through the support of Natural Law."[48]

—Maharishi

Maharishi once insisted, "Make the students fly," I answered, "I have to build a flying hall for them." It was too expensive after I built a school and maintained all the personnel. Later I realized how significant Yogic flying was in every aspect of administration.

Yogic flying removes conflicting tendencies and unify diversified qualities constantly occurring everywhere, among individuals, among countries and nations. There are many conflicts among governments and nations that deter the process of peaceful solutions. No wonder Maharishi was insistent and constantly refer to the program as a means to creating peace and a conflict-free world. It was the reason Maharishi wanted every nation in the world to have a group of Yogic flyers to finally unify conflicting tendencies constantly prevailing in the world. It was part of the Vedic norms.

---

[48] *Maharishi University of Management-Wholeness on the Move*

"In the vicinity of the phenomenon of Yoga, Yogic Flying, conflicting tendencies disappear. The unifying influence neutralizes the diversifying qualities in the area."

-Tat sannidhau vairatyagah—Yoga Sutra 2.35

Maharishi Staphatya Veda is a Vedic Science of designing the house or buildings in tuned with nature. The entrances are facing east and a central part is provided with an opening for the sun's energy twenty-four hours from sunrise to sun set. It enhances health and good fortune of the occupants. The neuron of the human brain reacts to the position of the sun. The east where the sun rises supports the firing of the neurons of the brain resulting in greater energy and more balanced functioning of the mind and body.

The blossoming of the sun flower or the morning glory flower that opens at a defined time in the early morning reflects the effect of the sun on earth as it sun rises from the east. The sunflower or the morning glory simultaneously blossom with the rising sun from the east. The human mind also blossom accordingly with nature with the rising of the sun. Man automatically sleeps at night when the sun has set and wakes up as the sun rises.

The TM - Sidhi program, the Yogic flying technique illustrate the Maharishi Effect, the mechanics of world peace.[49]

[49] *Book on Yogic Flying, Dr. Craig Pearson, Executive Vice President, Maharishi University of Management*

Scientific studies show that the neurons in the brain fire differently depending on what direction we are facing. The position of the sun with respect to the earth definitely affects the human brain and reacts to the circadian cycles of the sun reflecting the cosmic character of man. Maharishi strongly recommended that all residences and buildings be designed in accordance with natural law with the main entrances facing east or north. Based on the enfoldment of Vedic Staphatya architecture, Maharishi made plans to reconstruct the world thru the Maharishi Construction.

"Because the individual life is cosmic, everything about individual life should be in full harmony with Cosmic Life. Maharishi Vedic architecture gives dimensions, formulas, and orientations to the buildings that will provide cosmic harmony and support of nature to the individual for his peace, prosperity, and good health — daily life in accord with Natural Law, daily life in the evolutionary direction.

"Maharishi Vedic Architecture is the most ancient and complete system of architecture and planning according to the solar lunar and planetary influences on the earth with reference to the north and south poles and the equator – connecting individual life with Cosmic Life, individual intelligence with Cosmic Intelligence, creating ideal living conditions." --- *Maharishi Mahesh Yogi*

When archeologist found the lost city of Duarka, claimed to be one of the ancient cities that once flourished during the ancient Vedic civilization of India, found submerge beneath the Indian ocean, they discovered that all entrances were facing east. Many studies were conducted referring to east entrances and the other points of entry. It was found that east and north entrances are the most auspicious direction creating enlightenment, affluence and fulfillment. South entrances were found to be inauspicious creating negativities, problems and suffering. It is in contrast to the Chinese feng shui where they considered south auspicious. Maharishi strongly advised to move out of south entrance homes as if the house is burning. When the connection with inner intelligence is lost, symptoms of ill health begin to appear.

Maharishi Vedic Vibration Technology utilizes a refined impulse of Vedic sound, or Vedic vibration, to enliven the inner intelligence of the body and restore proper functioning of the human physiology. Maharishi Vedic vibration therapy soothes and enhances the senses and reduces chronic diseases. Maharishi Vedic Vibration Technology has brought relief to thousands suffering from chronic disorders, including people with arthritis, asthma, anxiety, back pain, depression, digestive problems, headache, and insomnia - all by means of a gentle, non-invasive vibration technology with no toxic side effects.

Maharishi explains that the body's own inner intelligence guides the proper functioning of a healthy physiology. Maharishi Vedic Vibration Technology describes the principles of transformation of the physiology from one condition to another are contained in the Veda and Vedic literature as specific sounds for specific parts of the physiology.

The underlying basis the of the human physiology are characterized by the vibrations of the finest particles of the human body. The fine frequencies described in the sounds of the Vedic texts balance the human physiology and targets the root, the non-material aspect of the physiology as they transform from fine non-material particles to matter forms.

The discovery of the Unified Field brought to light by modern science illustrates how the human mind and body is understood to be a waveform of faintest sounds, fine vibrations containing fundamental frequencies of an underlying non-material, unmanifest field, the Unified Field of natural law. It must mean that the super string theory is also at work within the humanphysiology.

When the frequency of the specific part of physiology is out of harmony, the body becomes sick and modern medicine has been unsuccessful to balance the underlying frequency of nature in that specific part of the body. Vedic Vibration can provide relief, sometimes instant by the use of the specific sound contained in the Maharishi Vibration technology which targets the non-material, unmanifest root of the disease. Using the proper sound and technology for the specific part of the physiology can instantly transforms disorderly functioning to orderly functioning of the body without invasive medical treatment. This mechanics of sounds are responsible to the transformation of natural law within the human physiology and responsible for the harmony and order in the functioning of the whole universe.

Maharishi's Veda and Vedic literature held knowledge of the fundamental faint frequencies of the body in accord with the natural frequency of natural law. The lack of understanding of these significant Vedic sounds has blocked the complete containment of chronic diseases. Those suffering from chronic diseases continue taking medications and still experience the symptoms of the diseases for many years because modern approach is incomplete and inadequate to restore health. The root cause of many chronic diseases are caused by the frequency imbalances inherent within the human physiology.

Tony Nader, MD, PhD is the author of the book, "The Human Physiology: Expressions of Veda and the Vedic Literature," and expressed the forty aspects of the Veda and Vedic literature are the same expressions contained in the human physiology. Dr. Nader is a brilliant neuroscientist and physiologist who worked closely with Maharishi who found the exact correlation between the structure of the 40 branches of the Veda and Vedic literature and the structure and functioning of the human body. He concluded that the only way to connect with the body and restore health is the use of the Vedic sounds establishing a scientific approach to modern medical treatment of chronic diseases.

Research shows that many people gain significant and lasting improvements, often within days. 'Instant relief' was reported by many people who were freed from their pain or symptoms within moments of beginning the TM program. Many researches were previously done and testimonies from medical doctors are available and anyone can apply at mvvt.org.

Maharishi's Gandharva Veda describes the music of nature and Vedic mathematics describe the sequential vibrations and the mathematical, non-material, unmanifest unfoldment of creation as it diversify into matter forms like planets revolving around the sun. Each of these melodies (Ragas) traces the vibrations and pattern of a particular time of day. When you play a melody during the specified time period, it creates a natural balance and harmony in your awareness and in the environment, and through the particular quality (Rasa) of each Raga, characteristics such as greater courage, self-confidence, wisdom, and happiness are enlivened.

Thousands of years ago in India, great Rishis or sages cognized within their own consciousness, the subtle rhythms and enlightening melodies of the Veda, which is pure knowledge at the basis of nature. As a part of the Veda, Gandharva Veda is the tradition of musical performance that replicates the vibrations of nature at different times of day and night.

Maharishi has revived this eternal music of nature and inspired the most renowned musicians of India to bless the world with Maharishi Gandharva Veda music. By playing this music you join in creating greater peace, harmony and joy for yourself, your dear ones and the world.

Direction	Gods	Representing	Planet	Part of the Building/Structure
EAST	**Indra**	**Lord of Lords**	**Ravi (Sun)**	**Entrance (Harbinger of good luck)**
Southeast	**Agni**	**Lord of Fire**	**Shukra (Venus)**	**Kitchen (Cheerful housewife)**
SOUTH	**Yama**	**Lord of Death**	**Angaraka (Mars)**	**Ghoulish Masks (To ward off evil)**
Southwest	**Niruti**	**Lord of Demons**	**Rahu (Dragon's Head)**	**Valuables (Prevents from being robbed)**
WEST	**Varuna**	**Lord of Water**	**Shani (Saturn)**	**Bathroom**
Northwest	**Vayu**	**Lord of Air**	**Chandra (Moon)**	**Bedroom (Serenity and calm and peace)**
NORTH	**Kubera**	**Lord of Wealth**	**Budha (Mercury)**	**Vault (Cash never runs out)**
Northeast	**Dharma**	**Righteousness**	**Guru (Jupiter)**	**Puja Rooms (For worship and meditation)**

VEDIC VASTU CHART[50]

The eight directions in Vastu Shastra show representations of the different ancient Vedic gods and planets. Accordingly Vastu determines the particular position of the different parts of a building or structure.

[50] *Stapathya Veda Texts*

Gandharva Veda is based on melodies called Ragas. Each Raga has a unique structure which combines both fixed elements and infinite possibilities of variation, allowing the musician to bring out all the subtle values of the frequencies present at that time of the day. Each Raga is based on one of the ten basic scales called Thaat, which determine the notes that are permitted in the Raga.

The knowledge of when to play each Raga comes from the time theory of Maharishi Gandharva Veda. This theory is based on three-hour periods called Praharas which correspond to the changing frequencies of nature throughout the day. It is best to listen to Maharishi Gandharva Veda in a comfortable position-sitting or lying down with the eyes closed. Even when no one is present to listen, playing the music 24 hours a day in your home or workplace generates a peaceful, soothing atmosphere that uplifts the whole environment.

In 1986, Maharishi Mahesh Yogi inaugurated his Program to Create World Peace, introducing the theme of creating balance in nature and peace in the world through the performance of Gandharva Veda music. Since then Gandharva Veda musicians have been performing in concerts around the year, around the world.

The Maharishi Ayurveda is most sought after by people with chronic diseases that modern medicine found impossible to completely eliminate. Panchakarma is that part of Ayurveda that has a long term effect produced by the removal of toxins in the body.

Modern food products particularly those products treated with preservatives or those that are genetically modified were found to cause incurable diseases such as cancer.

Maharishi's Vedic sciences are significant compilations from the ancient Vedic literatures that has now become important in our modern times. It is because these ancient technologies were forgotten and lost and Maharishi brought it back. Vedic sciences are competent technologies to solve the problems of modern man which targets man's problems at the root.

Outward manifest creation are more observable that the unmanifest root that is why it was missed and ignored by modern science. Like a tree, the leaves, the fruits and branches are seen outside the root. When the leaves are decaying, the leaves are investigated for the problem, not the root. It is the same thing with the problems of man. The diversity of man's environment has distracted modern science in uncovering the root cause of human failings.

The ancient Vedic sciences refer to the very root of man, inside his being. The recent discovery of the Unified Field of modern science brings out the root of nature occurring in the human mind. The Unified Field is enlivened at the deepest and quietest level of the mind. It is a great discovery in our modern times that can remove negativities in the life of man.

The Maharishi Effect is the ultimate solution to the problems of the world that has basis in the Unified Field, the very root of creation. Maharishi has brought to light the ancient Vedic technologies and gathered scientists to express the meanings of the ancient Vedic wisdom interpreted by modern science through the technology of the Unified Field.

Maharishi compiled the Vedic books of Patangili and used Patangili's yoga sutras in the TM-Sidhi program. Maharishi interpreted the Bhagavad Gita and worked with scientists to express their meanings into modern scientific format. Maharishi compiled the fragmented books of the Mahabarata, the Upanisads, Ramayana and the Puranas.

It is interesting to note that the Vedic literatures express the underlying unseen, non-material, unmanifest structure of creation also available in the human physiology and at the root of all the manifestations of the observable material creation. These brings to light that these ancient wisdom are the basis of modern science.

It astounds modern science, how an ancient civilization as far as 6,000 or maybe 12,000 years ago in antiquity were able to interpret the Vedic norms of life in tuned with natural law that is life beyond suffering. They have come up with literatures illustrating these norms, indicative that long, long ago, enlightened individuals ruled the earth in peace and prosperity. They have the knowledge of the unmanifest root of life while modern science had only explored the manifest, material aspect of creation.

Like the tree, the root has been missed by modern science. Maharishi has brought it back. His Transcendental Meditation targets that unmanifest level of reality occurring at the silent level of the human mind. It is now brought to light for modern man to use for its tremendous benefit.

Vedic science is that root. And because the answers to problems of modern man can be found at the root of nature occurring at the silent level of the human mind, Maharishi wanted every city, like Maharishi Vedic City in Fairfield in Iowa, to be the guardians of these precious knowledge of life beyond suffering.

Maharishi wanted these Vedic knowledge to be kept forever alive in modern times and in all the generations to come and never be lost again. Man need not suffer anymore. Vedic science brought forth the knowledge that can about bring heaven on earth.

It was a unique opportunity for our modern world to be blessed by the Vedic tradition of masters from the ancient Vedic civilization by the lifelong efforts of Maharishi Mahesh Yogi, the greatest of all masters. His works on Vedic science paved a new path for man to pursue peace, happiness and fulfillment in our generation. Otherwise suffering will continue on and on. Now we can look for a new transformed era that is suffering and problem-free. Studies of the Vedic sciences are now available at the Maharishi University of Management in Fairfield Iowa, USA.

Maharishi has brought to light Vedic systems of improving the life of man including Maharishi Jotish or Vedic Astrology that was once banned by some religion.

Maharishi Jotish or Vedic astrology has a significant effect in the individual's life because of the connection of man and the universe referring to the position of the planets at the time of birth. Maharishi Jyotish is the science of transformation and technology of prediction. It reveals the relationship of individual life with cosmic life and enlivens their fundamental basis in consciousness.

It is a practical program which helps one to "avert the danger that has not yet come" and to take maximum advantage of the fortunate periods coming in one's life.

Maharishi has said, "We are on the verge of creating Heaven on Earth, and therefore this knowledge of Jyotish blossoms now for everyone to enjoy heavenly consciousness—fully awake consciousness—reverberating blissful daily life. So here is an invitation to the people to create heaven in their life by developing their own consciousness through the study and practical programs of Maharishi Jyotish."

According to this traditional knowledge, prediction of future events is possible because the same orderly and sequential laws of nature which govern the evolution of the universe also govern the life of the individual.

Knowing any one point in the sequence, such as the time and place of birth, mathematically calculations can display the forwards and backwards in time that can assist in the evaluation of good and bad times in one life. If bad should appear, Maharishi Yagya can reverse the negative trends.

In this way the Maharishi Vedic Astrology program offers valuable predictive insights concerning tendencies in all areas of life -- health, partnerships, finances, education, career and family relationships. The Maharishi Jotish or astrology is like an alert system, this time or that time in any ones life, it can be seen like there is a light at the door.

A special feature of the Maharishi Vedic Astrology program is the availability of precise Vedic procedures, known as the Maharishi Yagya which help avert problems before they arise and to support good fortune in life. The Maharishi Yagya program helps to avoid future imbalances and is particularly necessary for anyone who is responsible for the well-being of others, such as parents, business managers, and teachers, because the destiny of others depends on their decisions.

Maharishi's Jyotish was further enhanced by Dr. Nader's discovery of the cosmic counterparts in the individual's brain. It scientifically confirms and identifies the participation of man in the functioning of natural law during the practice of Transcendental Meditation.

The human mind at the level of silence enlivens the seat of natural law, and contacts the Unified Field of all the laws of nature. With Maharishi's guidance, Dr. Tony Nader, MD, PhD interpreted the Vedic literatures as expressions of the human physiology including Tony Nader's second book on the *Ramayan in the Human Physiology* which is a discovery of the eternal reality in the structure and function of the human physiology.

"Maharaja Adhiraj Rajaraam, (Tony Nader, M.D., PhD), has discovered the precise one-to-one relationship between the fundamental structures and functions of human physiology and the fundamental structures of natural law. The fundamental structures of natural law connect individual intelligence with cosmic intelligence—the basic structures of the human nervous system with their "cosmic counterparts." *--Maharishi*

Maharishi repeatedly emphasized the importance and significance of Joyish, the Vedic astrology that guides the actual events that occur in any ones life, if the coming event is not good, it can be prevented before it occurs by the Maharishi Yagya program.

Every Maharishi Yagya performance brings more support of nature to the individual and enlivens natural law for the benefit of the individual or the whole society. The Maharishi Yagya service is also included in the Maharishi Vedic Approach to Health and Ayurveda programs for individuals with chronic disorders that is also offered at the Raj in Fairfield.

Traditionally, Jyotish has been used to diagnose problems, and yagyas have been used to help prevent them. The Maharishi Yagya program consists of Vedic performances by traditional Vedic pandits in India. According to this ancient knowledge, a Maharishi Yagya procedure, performed before the danger arises, can help neutralize negative influences so that they do not reach the individual, and life remains in tune with natural law. It is used to help avert undesirable circumstances or to enhance the success of an undertaking.

Dr. Tony Nader discovered the correlation of the nine planets to the human brain. He specialized in brain medicine and human physiology. In his second book, *Ramayan in Human Physiology*, Dr. Tony Nader described the story of Rama not just a cherished old epic of the ancient past. It actually illustrates the one-on-one connection of the characters, happenings and details and is being unfolded in every grain and fabric of the human physiology.

The ancient Vedic literature, the Ramayana, its texts, syllables, verses and the whole story is actually the Constitution of the Universe silently vibrating the language of natural law, the Unified Field administering the whole of the universe.

The human physiology is the Ramayana revealing the story of natural law unfolding within our brain and body, the cells and particles continually growing and incessantly moving like the vast universe forever expanding.

The illustration of the cosmic counterparts of human physiology, at the next page is from Professor Nader's book, *Human Physiology, Expression of Veda and the Vedic Literature*, presents a coronal section of the brain with its internal structures, and their one-to-one relationship with the nine 'cosmic counterparts,' the nine Grahas (planets). The diagram shows the sun, with the thalamus; the moon, the hypothalamus; Mars, the amygdala; Mercury, the subthalamus; Jupiter, the globus pallidus; Venus, substancia nigra; Saturn, putamen; ascending lunar node, the nucleus caudatus, head; Ketu or descending lunar node, nucleus caudatus, tail. The Sanskrit equivalent of the nine planets in sequence are Surya, Chandra, Mangal, Budha, Guru, Shukra, Shani, Rahu and Ketu.

As the ancient Rishis say, "As within, so is outside in the universe." Natural law is functioning inside the human physiology as it is functioning in perfect order and harmony in the universe.

Dr. Tony Nader displays the long sought after principles of life in harmony in accord with the orderly functioning of the universe to be as orderly down on earth. It defines that life on earth is meant to be orderly and in harmony that is beyond suffering, beyond poverty and disease. Climate or air turbulence, wars and all negativities all balance creating harmony inside man and the whole of his environment. Man is truly cosmic whose birth right is happiness man was meant to be.

Maharishi recognized that the ancient Veda and the Vedic literature held the secrets to man's fulfillment and therefore gathered them piece by piece from obscurity and worked with scientists of today to interpret them into modern language. Modern science can now make use of these discoveries to perfect medical and health sciences, to improve the science of psychology and rehabilitation or further uplift other manifest sciences above the level of the Unified Field. Dr. Tony Nader's discovery completes and brings to fulfillment Maharishi Vedic science. He has become the most brilliant scientist of modern times confirming that the ancient Vedic literatures are truly the expressions of the Veda containing the total knowledge and functioning of natural law within the human physiology. The total knowledge of natural law has now been brought to light and becomes available in each and every human being.

Dr. Tony Nader's brilliant achievement represents one of the most outstanding and significant discovery in the history of mankind. He has identified the merging point of the oldest ancient traditions of knowledge held by the ancient tradition of Vedic masters. Dr. Tony Nader has become the greatest scientist in modern times following in line with the ancient tradition of masters and his name will be revered for many generations to come. He has ultimately interpreted the ancient Vedic science into modern science in understanding man's great potential and means for the creation of heaven on earth.

The Vedic norms from the ancient Vedic civilization are very significant in modern times because man suffers because of violations of the Vedic norms, which are violations of the norms of natural law.

These norms were expressed in many world religions like contained in Christianity's Ten Commandments. In the fourth commandment in the Christian bible, it states, "Honor thy father and they mother." Maharishi exemplified his service to his teacher by his lifetime devotion, honoring his master Guru Dev. In all his lectures, he ends, "In honor of my teacher, I bow down at the feet of Guru Dev. Jai Guru Dev."

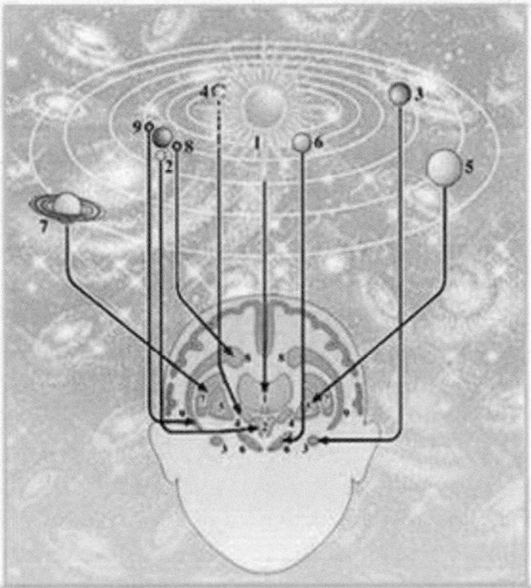

① • Surya or Sun ① • Thalamus	④ • Budha or Mercury ④ • Subthalamus	⑦ • Shani or Saturn ⑦ • Putamen
② • Chandra or Moon ② • Hypothalamus	⑤ • Guru or Jupiter ⑤ • Globus Pallidus	⑧ • Rahu or Ascending Lunar Node ⑧ • Nucleus Caudatus, head
③ • Mangala or Mars ③ • Amygdala	⑥ • Shukra or Venus ⑥ • Substancia Nigra	⑨ • Ketu or Descending Lunar Node ⑨ • Nucleus Caudatus, tail

COSMIC COUNTERPARTS OF HUMAN PHYSIOLOGY[51]

In our modern times, many miss this wisdom of honoring people who have provided support sometime in their life, like their parents. By honoring them, they give honor to themselves, otherwise their life becomes incomplete because they fail to live in accord natural law. Parents first than friends, this is nature's norm.

Some children leave their parent alone during special days like Thanksgiving day or Christmas time and go with their friends to some mountain climbing in Peru or go somewhere else. They do not feel that their parents value their presence even only on special days. It is tragic that families now leave their aging parents in homes for other people to care for them. The greatest joy of parents is their

---

[51] *Human Physiology, Expression of the Veda and the Vedic Literatures by Tony Nader, MD, PhD*

children when they were young and to see them occasionally when they are old. We must not intentionally withhold this joy from them. It is a violation of natural law and violation of natural law brings cause and effect.

Some do not value service like Maharishi's lifetime service to Guru Dev, his teacher. The natural law norm of service obliterates the ego that blocks the flow of the Unified Field in their lives. Service of a wife to her husband or vice versa is of great significance. Many modern marriages break up, not knowing this secret knowledge of service. Service is the greatest of all virtue, greater than love. Service brings so much support of nature to the one that serves. It is the servant that creates the magic of fulfillment to himself, not to the master. Service obliterates the ego that block of the unfolding of the fullness of natural law in the life of every one.

It is like Maharishi's service to his master Guru Dev. Maharishi was able to unfold the totality of natural law available in the inner most silent level of the mind that allows everyone to live in accord with the norms of natural law within man.

These norms are vital in the life of man because of the support of nature that comes naturally with it or does not come without it. The Vedas say, "He who serves a master gains total knowledge." Maharishi was an embodiment of the Vedic norms that he was teaching for fifty years. Even in his death bed, he said, "I have completed my duty to Guru Dev."

No man in the history of the world has come up with a technology to completely wipe out suffering and create heaven on earth. With service and devotion to his master Guru Dev, Maharishi unfolded the knowledge of how to create heaven on earth.

Maharishi's practice of Transcendental Meditation naturally transforms man into an intelligent being, a compassionate being and an enlightened being. With Transcendental Meditation, man becomes in tuned with the will of God, in tuned with the laws of nature. Man does not violate the commandments of God. Automatically, with the practice of Transcendental Meditation, man becomes happy, kind and compassionate.

In the history of time, many norms to change the course of mankind's destiny has come and gone. They were unable to create the change for the good of humanity. Religions evolved generation after generation, but they were unable to transform man and unable to create peace, on the contrary, throughout history, many wars were fought in the name of religions, in violation to the commandment

"Thou shall not kill." Killing still goes on. We see these wars occurring in the Muslim Middle East where Christians are being displaced or killed. Maharishi's Transcendental Meditation is not a religion but a technique of quieting the mind to reach the Unified Field prevailing at the quiet level of the human mind.

Maharishi's Transcendental Meditation makes man happy, a happy man does not get a gun and shoot another. Many aspects of Maharishi's Transcendental Meditation are expressed in many religions. Like when Buddha was asked, "What have you gained from meditation?" He replied, "Nothing." However, Buddha said, "Let me tell you what I have lost: Anger, anxiety, depression, insecurity, fear of old age and death."

For the past twenty centuries, religions have spoken about this great wisdom within man, the God within. Even Jesus Christ said, "Seek ye first the kingdom of God within and all these things shall come unto you." (Matthew 6:33). "The kingdom of heaven is within you." (Luke 17: 21). "Where two or three are gathered in my name, I am in the midst of them." Natural law principle of economy was expressed in these verses including the group effect, the Maharishi Effect. The will of God is expressed thru natural law.

Plato and Aristotle, who were great thinkers and philosophers of previous centuries, they spoke of the inner silence within man. St Augustine said, "I entered into the innermost part of myself." St. Gregory the Great, an heir to a large Roman family fortune decided to become a monk and became a Pope, he wrote, "The mind is frequently carried away into the sweetness of heavenly contemplation- it sees something of the inmost realities as if it were through the mist." St. Theresa of Avila wrote, "My soul at once became recollected and I enter the state of quiet rapture."

Thomas Merton, a learned graduate from Columbia University and a famous American monk wrote, "The utter simplicity and awareness of the infused light pours into our soul suddenly over comes us to a new level of awareness."

Most of them have enjoyed contemplative silence, not like living in our most modern life with TV's, internet and modern gadgets that keep the mind engrossed in the fast pace of modern living. Silence now seems to be unimportant and seldom available.

One need not become cloistered. In the fast pace of modern life, TM can bring the silence that anyone can enjoy any place, anywhere. The path to God is easy. God consciousness is available to all religions. What is lacking is the level of silence in the human mind that Transcendental Meditation can provide.

In Buddhism, it is called Nirvana. In Islam, it is Fana. In the Indian mythology, it is called Yoga. Maharishi, Einstein, and Hawking call it the Unified Field of modern science.

Philosophers, theologians and scientists have spoken of the same state of transcendence experienced in the inner most sanctum of the self. I call it the field of the anima solo particles, the finest most powerful particle of nature, the unseen "sap" within every creation.

Call it by any name it is the same banana, ancient, past, and present. The future of science will produce a name, but it will still be the same banana describing the qualities and attributes drawn by the quiet level of the mind. The fruit is determined by its seed. What causes it to blossom and bear fruit is determined how well the seed and its root is nourished. Transcendental Meditation nourishes the root of man and naturally blossom in fulfillment.

Transcendental Meditation Study on the Effectiveness in Reducing Trait Anxiety Meta Analysis

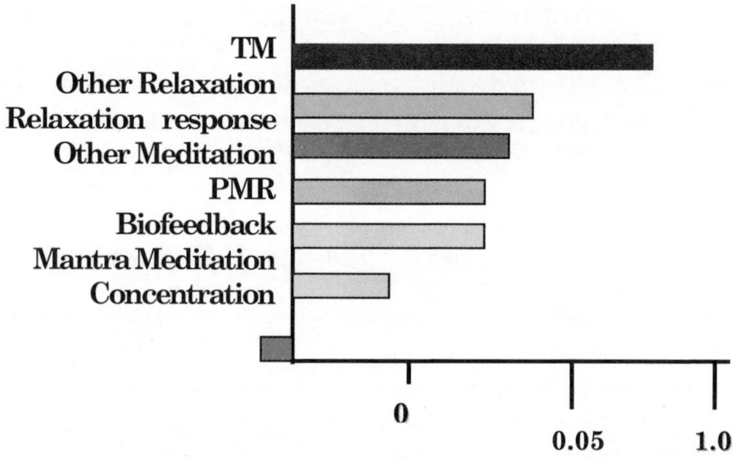

Journal of Clinical Psychology 1989; 45: 957-974

REDUCTION OF ANXIETY

.Most scientists do not believe in God. Maharishi as a scientist and physicist expounded God consciousness as the evolutionary path to God and believed that there is a God that created order and design in the universe. Maharishi wanted all men to experience the spirit and the will of God easily thru the practice of the Transcendental Meditation technique that he brought down out into the world from the deep forest of the Himalayan mountains.

Isaac Newton believed that the solar system could not have been created without a God. Stephen Hawking believes there is no God and challenged Newton's theory of God and referenced God as a force (or maybe the force field described in Maharishi's Unified Field) that could explain the creation of the universe.

Stephen Hawking stated "That God is defined as the embodiment of the laws of nature and because there is a law such as gravity, the universe can and will create itself from nothing, spontaneous creation is the reason there is something rather than nothing. That is why the universe exists, why we exists. It is not necessary to invoke a God to set the universe going."

The merging of science and religion meet at the silent level of the mind at the level of the Unified Field. It is the seed form of natural law and the flat form of the will of God. Transcendental Meditation automatically transforms man, illustrated by many scientific validations from distinguished scholars and from prestigious scientific journals.

The results of these validations cannot anymore be ignored. They are irresistible to the intelligent mind. There are now many intelligent men that have risen in our time. Many scientists have come up to validate Maharishi's ancient Vedic science, the Veda, that contains silent fine frequencies that expresses and unfolds the norms of life naturally occurring without the use of any man-made law or commandments. The Vedas contains all the laws of nature and its seed form that can be harnessed by the mind at the level of silence, at the level of the Unified Field bringing forth the total transformation of man and society.

The Unified Field has been found to transform man and his environment. Society has now a technology to transform itself from poverty to prosperity, from war to peace and for having balanced climates without wild storms, fires and destructive tornadoes.

Truly God did not create man to suffer, God provided man a tool that was kept by the ancient Vedic culture claimed to be the oldest civilization in the world. These ancient Vedic norms will restore the will of God on earth.

Maharishi Effect described the Rig Veda, it is the most significant discovery of modern times to transform man into his highest status of dignity man was destined to be. Once the 9,000 group of TM and TM-Sidhi participants are established in one place, heaven on earth is not very far away into the future. Fairfield has that seed that has survive more than forty years.

Maharishi's achievements extended fifty years and brought Transcendental Meditation to the attention and practice of more than six million world population from 192 countries as far as Brazil to Ukraine. The reason why wars still are waging and storms are still prevailing, the 9,000 has not been established in any place in the world. Two thousand are at the central part of India and one thousand in Fairfield, Iowa, USA. Continuous effort is being exerted by the Global Country of World Peace throughout the world.

Transcendental Meditation has reached top level of individuals the United States. Many celebrities and prominent individuals embraced Transcendental Meditation because of the inner peace experienced brought about by the practice of Transcendental Meditation easily provides with one's self and with the world. There is truly something that eludes complete enjoyment of external success. Transcendental Meditation provides that stability, improved health, inner peace, and deeper happiness outside that of fame and fortune.

20

NATURAL LAW PRINCIPLE OF CREATING GLOBAL WEALTH

Maharishi said that the success of growing a tree is watering the root. The success of governing a country is applying the Unified Field, the seed form of natural law found lingering within the silent level of the human mind in large groups together. The Unified Field of natural law waters all life. For the world, it is the square root of one percent of the population practicing Transcendental Meditation and Yogic flying in large groups together. For the world now, it is equal to 9,000. Countries with smaller population, it is just a few hundreds. Small nations in poverty can avail of this newly discovered natural law principle of creating global wealth as it needs just small number of their population to practice TM and Yogic flying in groups together twice a day, morning and afternoon. TM quiets and allows the human mind to draw the qualities of the Unified Field that waters the root of life that unfolds peace and prosperity to prevail in the community, country or the world.

After the Maharishi University of Management in Fairfield, Iowa was established in 1974, Maharishi invited practitioners of Transcendental Meditation from all around the country and the world to relocate and gather in Fairfield to join the group practice of TM and Yogic flying program to create the daily super-radiance. The super-radiance effect of the Unified Field produces positive trends in the environment.

The Cambridge Investment Research was one of the companies that relocated in Fairfield in 1982. Soon after being based in Fairfield, the company's income jumped from $500,000 to a hundred-fold, to $50 million dollars[52] in 2002. It described the phenomenon similar to the ten-fold trend of increase of the stock market in just three weeks during the 1983 Global Taste of Utopia where 7,000 TM and Yogic flying participants gathered in Fairfield described in Chapter 6 and that is illustrated on Chart 6 and 7.

In June 30, 2013, the Cambridge Investment Research company garnered $53.4 Billion dollars of managed funds and an equivalent gross income of nearly a thousand-fold increase equal to

[52] *The Washington Post August 9, 2004; Page A03*

$461.4 million dollars.[53] In May 2010, Eric Schwartz, Founder and CEO of Cambridge Investment Research was honored by Investment Advisor magazine as one of the thirty most influential people in business over the past 30 years. He credits his success to the practice of Transcendental Meditation that he learned when he was freshman in college.

The Cambridge phenomenon may not be just a coincidental occurrence. It is a scientific phenomenon, when groups daily practice the TM and the TM-Sidhi program in large groups together, in one place, a super-radiance effect results in the surrounding environment. Like in all experiments conducted worldwide, the same phenomenon occurs. It is the same result in what Maharishi said about a tree, water the root, the leaves turn green, it blossoms with flowers and bears fruits. Maharishi refers to the increase in profits and improved economy to natural law principle of creating global wealth, like a tree bearing fruits with the Unified Field watering the roots. The group practice of TM and Yogic flying must have watered the root of the Cambridge company that increased its profits thousand-fold in Fairfield, Iowa where there are large groups of practitioners of TM and Yogic flying daily. Fairfield in Iowa, USA is just a small city. If every small country form their own group, many success will be produced and the country will experience economic rise, just like in Mozambique, Africa in 1994, when large groups of the military practice TM, the economic index changed from the long prevailing poverty to 19% rise in economic growth, including the cessation of hostilities that grappled the country with nearly two decades of civil war. Abundant rain simultaneously fell in the long drought stricken country of Mozambique.

In the schools which I had built in the Philippines, it displayed the same results, property value increased nearly a hundred-fold. It raised the economic status of the remote poor town as soon as the students practiced Transcendental Meditation in large groups together.

On the other hand, the success reached by some billionaires and millionaires in the USA might be because of their Transcendental Meditation practice. It may appear that Ramen Ayer, former CEO of Hartford, Eric Schwartz, CEO of Cambridge Investment Research in Fairfield, Iowa or Ray Dalio, CEO of Bridgewater Associates, the largest hedge fund in the world, all billionaires who have earned their fortunes because in their youth they learned TM and practiced

---

[53] *CIR2 Website, Fast Facts 2013*

Transcendental Meditation in the past for nearly four decades, their TM practice seemed to have contributed to their success.

Ramen Ayer says of TM, "I have been practicing the Transcendental Meditation technique for over 30 years. Transcendental Meditation has helped me to consistently raise my performance levels. It has helped me cope with the stresses and strains of life. And I believe very strongly that the practice of meditation has helped me to maintain a steady state of mind – it has given me equanimity no matter what happens."

Eric Schwartz learned TM in college and then young Ray Dalio followed Maharishi in India and learned TM along with the Beatles. After more than forty years, both Eric Schwartz and Ray Dalio have become among America's great men of fortunes and they refer to Transcendental Meditation as part of their success.

I have been doing it for 44 years, twice a day for twenty minutes," says Ray Dalio, the guru behind Bridgewater Associates, "It's such a great investment, more than any factor in my success. It opens up the two sides of the brain, brings creativity and open-mindedness. It allows you to clear your head and bring an equanimity to everything. Meditation more than anything in my life, was the biggest ingredient of whatever success I've had."

"There are different parts of our brain that have a very big influence on us. The amygdala is the part of the brain that has the "fight or flight" that produces anxiety. The pre-frontal cortex is the part of the brain which we call "for executive functioning," where it's the part that's reflective. It's the part that we are calm, and we say, "Do we want to do that? Do we not want to do that?" We put things in perspective. Those two parts of the brain fight with each other. In other words, the passion: "I'm going to do that because it's exciting but it may be harmful." The other part of the brain says, "You don't want to do that. So that whole emotional fight-or-flight part of the brain during meditation, through brain imaging they see that calms down, and the pre-frontal cortex lights up. So that open-mindedness creates where creativity comes from. Because creativity is not coming from the "working the brain" and the "I will work hard and think about it" and that "I will muscle it through." It comes from this relaxation. It's an opening up, and takes a hot shower and don't be thinking of something and some great idea comes through and you grab the great idea."

"So meditation is very much like that. It opens the mind. It creates an openness, a freedom, in which – I don't know whether we would say "an intuition" – that creativity just kind of comes through and it creates an equanimity. In other words, you could step back and you can put things in perspective. It doesn't lessen your emotions. The emotions are the same, but you can step back and say, "I'm not going to be controlled by that emotion," or "Let me put things…" and I think it then helps to see things at a higher level."

Ray Dalio's meditation regimen has inspired other Wall Street celebrities, including Daniel Loeb and New York Times columnist Andrew Ross Sorkin, who made the interview said he meditated the night before conference to calm his nerves.[54]

Ray Dalio has publicly declared that Transcendental Meditation has a significant part in his success. He started with a $300 dollar investment he saved working as a golf caddy as young student. His company, the Bridgewater Associates has become the largest hedge fund company in the world. In his drawing board is a new $750 million headquarters Dalio is working on to build in Stamford, Connecticut, to accommodate more than 2,000 employees, nearly double the number of people currently working at his firm.

In February 11, 2014, in a New York town hall, Ray Dalio, Dr. Memet Oz, Mario Batali - chef, writer and restaurateur and Andrew Ross Sorkin - CNBC's signature morning program, Squawk Box, all practitioners of TM, together they participated in a forum for leaders on the research and application of the evidence-based Transcendental Meditation for promoting creativity, enhancing performance and overcoming stress. The forum was sponsored by the David Lynch foundation.

The rich in the country hold in their hands the ability to maintain the status of the USA as a permanent economic super power that seems to be slipping with the huge increase of its trillions of dollars in national debt. The rich can reverse the rumors of the American dollar falling or that US economy is at the brink of another economic fallout. Prophets of doom predict the next economic collapse will occur larger than the 2008 recession and without immediate recourse. Of course there is arecourse for the US government, it can look at the permanent groups in Fairfield that practice twice daily the TM and Yogic flying. Fairfield needs to increase the number of participants to 2,500, the sustainable number in order to make the USA invincible from negative trends. The USA has the largest practitioners of TM in the world, nearly one million all scattered in the country, it may be helping the USA maintain its economic stability against many odds.

---

[54] *IB Times, November 12, 2013*

The rich have hundreds or thousands of employees and they have facilities to house practitioners of TM and the TM-Sidhi program creating the Maharishi Effect that can avert any eminent economic collapse to occur. Recessions and economic downtrends can be prevented by the large groups practice of TM and the Yogic flying together in one place evidenced by several large events conducted around the world. In Fairfield, in 1984 world economy showed an instant rise, with major stock markets of the world rose during the Global Taste of Utopia event illustrated in Chapter 6 shown on Chart 6 and Chart 7.

Currently in Fairfield, Iowa, USA, about one thousand Yogic flying participants have already been established by the residents and the students at the Maharishi University of Management. At the central part of India, already facilities have been built to house two thousand Vedic pundit boys. The rich can financially support these already on-going efforts.

Celebrities spearhead the rapid expansion of Transcendental Meditation around the world. In the early 1960's at that time, the popularity of the Beatles came about rapidly as they followed Maharishi and they learned the Transcendental Meditation technique. The Beatles proclaimed to the world the benefits of Transcendental Meditation during that era when they were the most sought after rock stars in the world.

Martin Scorsese, film director, George Lucas and David Lynch are famous film producers who have learned TM. David Lynch formed the David Lynch Foundation assisting inner city youths to learn TM and introduced TM to veterans with post traumatic disorders. The Transcendental Meditation program has helped those who were at higher risk for post-traumatic stress in the military.

Thru the David Lynch foundation, TM was taught in schools where students grow up in a climate of fear with bullying, violence and substance abuse. The David Lynch foundation has also worked with other at-risk populations like Native Americans and the homeless. The foundation sponsored TM to nearly 150,000 students in 350 schools around the world, fifteen of them are located in the United States.

"It's not a religion," Lynch said, after an interview by ABC-TV. "It's not against any religion, it's not mumbo-jumbo. It truly does transform life. Kids come to school and they meditate together for 15 minutes in the morning. And before they go home they meditate for 15 minutes. A lot of them come from, you know, bad situations, and so this gives them this thing you know, at the beginning and the end of the day, the rest of the time you just watch the violence stop. Watch relationships improve. Watch happiness in the hallways, in the classroom, watch creativity flow more and more, watch that heavy weight that we are living under gently lift away."

Various studies funded by the David Lynch Foundation have shown that those under stress, particularly ethnic and racial minorities, can reduce their stress levels by 36 percent by practicing TM. Students in "quiet time programs" that include Transcendental Meditation, have also shown higher rates of achievements. These groups need constant supervision for results to become permanent.

Many billionaires and millionaires in the USA practice Transcendental Meditation. Rupert Murdoch is into Transcendental Meditation. The Australian-American media magnate was born in 1931, is an Australian American media magnate. He is the founder, Chairman and CEO of News Corporation, the world's second-largest media conglomerate. News Corp includes television, movies, cable networks, book publishing, satellite TV, magazines and newspapers operating on every continent. Murdoch's best known brands include 20th Century Fox, Fox Television, DirecTV, Harper Collins, and MySpace. Murdoch is the 38th richest person in the US and the 106th-richest person in the world, with a net worth of $8.3 billion. He has been listed three times in the *Time* 100 as among the most influential people in the world.

Famous radio personality Howard Stern learned Transcendental Meditation and has been practicing TM for many years. Several individuals followed him and learned TM after he spoke about the many benefits of TM one time in his radio talk show, Xiruis XM.

Comic actor Russell Brand credits his sobriety with practicing Transcendental Meditation. Russell Brand, who credits Transcendental Meditation for helping him stay off addictions to alcohol, drugs and sex, did a stand-up comedy show at the Palace of the Fine Arts in San Francisco to benefit the David Lynch Foundation. Brand has said publicly that meditation had helped him find a "deeper state of happiness."

Many celebrities, including Oprah Winfrey and Ellen DeGeneres practice TM. Oprah Winfrey practice Transcendental Meditation including four hundred of her staff and featured Maharishi Vedic City and Fairfield in her OWN-TV show and because of this show, hundreds of people took to learn TM after she aired Transcendental Meditation in her show.

Transcendental Meditation has gone main stream in America because people of prominence declare to the public the enormous benefits of TM backed by numerous research data that supports their endorsements. Oprah Winfrey truly represented a gigantic endorsements that many people benefited from. The general public can access the stories of celebrities with the aid of TV, the press, general mass media and the internet at www.tm.org.

Dr. Memet Oz, including his staff practice Transcendental Meditation and stated that TM can reduce high blood pressure better than some blood pressure reducing drugs. It was found that TM is the only system of cure that eliminates the cause without any harmful side effect.

Dr. David Orme Johnson, PhD, one of many scientists who has done extensive research on Transcendental Meditation. He published his works in several scientific journals and thoroughly described them at www.TruthAboutTM.com, the website he has created for the purpose of illustrating the scientific base of TM and Yogic flying. His work encompasses one of the most extensive researches on Transcendental Meditation and the Maharishi Effect. He has travelled to over 56 countries speaking about TM and its scientific base to different scientific conferences, to the public, the press, members of governments including the United Nations.

CNN Anchor, Candy Crowley, is the first woman in twenty years to moderate a presidential debate, that was between Mitt Romney and President Barack Obama in October 19, 2012. She is a practitioner of TM. She was the commencement address speaker at the 38th commencement graduation of students from 37 countries at the Maharishi University of Management in May 26, 2012. She has supported in the teaching of TM to inner city youths and war veterans with post traumatic stress diseases.

Students, faculty and officers of the Maharishi University of Management applauded Senator Tom Harkin of Iowa, in his commencement address to the graduating class of May 25, 2013 when he praised the Maharishi University of Management as "the best holistic approach to education and wellness in life of any university anywhere on the globe."

People of prominence greatly influence governments and the general population and because they belong to that great level of intelligence and success, Maharishi has called successful individuals, "friends of natural law." The general population look up to successful people as role models and follow their trends that is why their role of influence provides a significant and far-reaching impact in the spreading of the enormous benefits of Transcendental meditation and Maharishi's invincibility programs.

The TM community feel honored to read endorsements in the press and media from prominent people doing their share of spreading Maharishi's supreme technologies of natural law to the general public. It has now become easier to talk about TM not being a religion because prominent people say so. Several decades ago, it was an effort to convince people that TM is not some kind of an esoteric mumbo-jumbo practice. TM has become part of a modern science discovery, proponents of its scientific base was uncovered by Dr. Hagelin, PhD; David Orme Johnson, PhD, Robert Keith Wallace, MD, PhD and several other distinguished scientists. They have brought forward many validated research supporting the endorsements of people of prominence. Many of the scientists and researchers are recognized by the scientific community coming from even the prestigious National Institute of Health of the USA which has provided nearly $24 million dollars of funds for the research.

Maharishi's successor Maharaja Adhiraj Rajaraam, Tony Nader, MD, PhD, now first Ruler of the Global Country of World Peace, himself a distinguished and brilliant scientist who discovered that the ancient Vedic literatures referred to by Maharishi, are text and sounds, fluctuations of natural law traversing the whole human physiology and enlivening the self-healing and nourishing character of nature built in within every human being. He found and confirmed what Maharishi has said that the sounds and vibrations occurring in the human physiology parallel the subtle vibrations of natural law flowing in the whole vast universe.

In the ancient Veda, it said, "I am Brahm, I am totality." The Vedic verse describes the cosmic character of man. The Unified Field of modern science unfolded the existence of the finest particles lingering within the human mind at the level of silence during the practice of Transcendental Meditation.

It took nearly fifty years for Maharishi's technologies to become well known all over the world. Unlike Galileo of the sixteenth century, a physicist, mathematician and astronomer who discovered that the earth was round. It took nearly sixteen hundred years before the earth was accepted by the world to be round not flat. Galileo played a major role in the scientific revolution reversing Ptolemy's flat earth theory in 100 A.D that even Plato, Aristotle, ancient Greek philosophers including the Church believed Ptolemy's theory to be true.

Maharishi has played a major and supreme role in the enfoldment of a technology to transform humanity to the status of happiness and fulfillment reversing that problems and suffering are part of humanity. Maharishi has proclaimed that heaven can be created on earth through the group practice of TM and the TM-Sidhi or Yogic flying.

In just fifty years, modern scientists have interpreted Maharishi's ancient Vedic science and brought to light the Unified Field expressing the invincible field that shields the earth from any negative trends than can further nourish life on earth for all times.

Several private groups have taken into TM and Yogic flying and Maharishi made an urgent invitation for governments to embrace it. Hopping is the first step to flying, then hovering and finally when collective stress is completely eliminated and indomitable bliss start to radiate in the whole world, flying is expected occur.

Flying is illustrated when the human brain reaches its maximum brain wave coherence, the body lifts up defying gravity. The most fundamental level of silence in human consciousness where all the four fundamental forces of nature, electromagnetism, weak force, strong force and gravity meet and unite, it was found to occur within the human mind at the level of silence.

In Fairfield, Iowa, USA about 1,000 are hopping blissfully daily twice a day, all that is needed is to increase the number to 2,500 to gain permanent prosperity and invincibility for the whole of the USA.

In the central part of India, about 1,300, Vedic pundits are performing Yogic flying. In Central America, in Monte Alban in Mexico, in Thailand, in Africa, in Skemersdale, England and many other parts around the world, pockets groups are daily practicing Maharishi's Transcendental Meditation and Yogic flying.

Soon there will be a time when war will be happening less and less in the world than today. Even Russia's aggression and annexation of Crimea anxiously watch by the whole world, was predicted to be a start of another world war after twenty years of peace between Russia and the Western world. The USA and the West exercised restraint preventing military intervention instead sanctions were exchanged by the West, exchanging military arms hopefully will not happen. A Transcendental Meditation center exists in Russia, www.Maharishi-TM.ru. TM and Yogic flying can stabilize Russia's failing economy due to the fall of oil cost and due to Western sunctions.

Even Iran has agreed to temporarily reduce their nuclear program in exchange for relief from sanctions and maybe permanent accord will be signed in the near future. This is the first peaceful agreement signed in over thirty years with Iran.

In the Middle East, the Muslim religion, its interpretations made political by different diverse groups, they have been fighting since biblical times and nothing has united their different ideologies into one cohesive endeavor for living among themselves in lasting peace. In the event 9,000 will be created in one place on a permanent basis, heaven will be established on earth. When this event happens, the establishment of a permanent large group, all countries will be prosperous and be at peace. Any intent of aggression and conflict will melt away.

When the 9,000 TM or Yogic flying participants will be in place on a permanent basis, or a square root of Syria's population, the war in Syria will end where some 250,000 people have been killed, communities and infrastructure destroyed and millions of refugees fleeing Syria, some living under deplorable conditions along border countries, they will all come home.

Conflicts in Syria, Yemen, Iraq, Libya, Afghanistan, Israel and Palestine need a group of Yogic flyers to solve their problems on a permanent basis. All wars will end and never will it rise again. Life on earth will rise to the level of supreme status of fulfillment man was destined to be. Enlivening natural law within the human mind at the level of silence during the practice of Transcendental Meditation is the most significant discovery of the twenty first century because suffering and problems man has encountered for many, many centuries will come to an end.

When the organizing intelligence of natural law will be embraced by governments, whatever form, democratic, parliamentary even communist form of government, nations will rise to invincibility against negative trends and permanent prosperity will prevail. The rise of invincibility, starts with the practice of yoga, of unity with all life, the integration of the totality of life thru the practice of Maharishi's Transcendental Meditation opening awareness to bliss consciousness, to the Unified Field of all the laws of nature.

No negative trends can exist in the Unified Field, like in the magnetic field, any impurity will be thrown out of the magnetic field. It is the character of the Unified Field to shield off any negative trends in its environment. It is a natural phenomenon in nature. The electric field is created by the electron particles, the magneto particles create the magnetic field, the gravitons create the gravitational field.

Friction resists the flow of electricity in the copper wire as the human Ego resists the flow of the Unified Field blocking fulfillment of desire. The Unified Field holds the finest of all particles beyond electrons, magnetos and gravitons. The AS_U particle that animates the whole universe lingers within human awareness creating the invincible field, the Unified Field prevailing in the whole universe.

The Unified Field is the field of unity in nature where all the four fundamental forces of nature of nature, electro-magnetism, weak force, strong force and gravity unite and become the seed form of natural law.

Maharishi's program of Yogic flying illustrate the point of unity in nature, where the body lifts up at the point of maximum brain wave coherence defying gravity. This point of unity, the point of yoga describes full integration of life where only positive trends prevail. Yogic flying eliminates negative tendencies that results in conflicts and unifies diversified influence in the surrounding environment. A large group of Yogic flying participants create an invincible field from negativities occurring including providing the balance of temperatures that controls climate changes.

21

COUNTERING TERROR AND CREATING WORLD PEACE

Nearly seventy years ago, after World War 11, the United Nations was established with the purpose of preventing the recurrence of war. Yet many wars have occurred until now, like in Iraq, Afghanistan, Yemen, Syria and several parts of Africa.

All wars, like global warming are man-made. Global warming occurs when the greenhouse layer in the atmosphere overloads with CO_2 that is caused by man - made manufacture and use of fossil fuels, which results in greater heat absorbed that is reflected from the surface of the earth which create droughts, the rising of sea levels and the increase in ocean temperatures that cause floods, typhoons and hurricanes on earth.

Maharishi has said that when collective stress overloads in a country, conflicts rise and war breaks out. Wars and conflicts are man-made concerns. In the individual, when stress overloads, it can cause disastrous diseases like high blood pressure. Transcendental Meditation releases stress. Dr. Memet Oz, a prominent American doctor has said that TM can reduce and relieve high blood pressure better than some prescription drugs for high blood pressure without side effects.

The practice of TM and Yogic flying established in large groups together can solve conflicts and war better than bombs, missiles or any military arms. The explanation is described by the properties of the finest elementary particles of nature occurring within the human mind at the level of silence. Referring to the finest particles, without any name available to date, enlivened within the human mind, I referred to it as the ASu particle, the particle that animates and nourishes the whole universe. Maharishi referred to it as the "sap" of the universe that creates "by itself, from itself and within itself." Scientist called it the Unified Field. These particles traverse every single cell of the human nervous system and remove the negativities that cause imbalance in the body, then nourishes and heals the body. Several scientific research now show many health benefits of Transcendental Meditation.

The scientific facts displayed by fine particles of nature indicate that they hold exponentially powerful character in group form which create a field, like the laser or the photon field. One photon particle creates insignificant effect than when they behave in groups together, they create an exponentially powerful field of light which spreads over a specific area. No darkness can exist in a photon field. Like in a magnetic field, the magneto particles when joined with More magnets together create a greater magnetic field that becomes invincible from impurities that come into its field. The strength of fine particles of nature in group form becomes exponentially powerful field described in quantum physics. Like the hydrogen particle in atomic form, it can create a huge distractive blast when concentrated in tons. The Unified Field, the finest particle of nature enlivened within the silent level of the human mind creates an invincible field where no negativity can exists. It cannot create destruction, rather, its natural character is to nourishes life. Like a tree, water the root and the whole tree is nourished. The tree grows, blossoms and bears fruit. The Unified Field nourishes all creation. These are the reason all negativities in the surroundings like conflicts, terrors and war fade away as soon as its invincible shield is spread over a specified area covered by the square root of one percent of the population radiating coherence within the area. Existing evidences already exits and to verify the fact, it can easily be repeated again and it will show the same positive results in the environment.

The Global Taste of Utopia assembly called for by Maharishi in 1983, in Fairfield, Iowa, USA, it showed to the world that the 7,000 TM and Yogic flying participants in group form, collectively gathered in Fairfield resulted in the great reduction and the cessation of world conflicts as shown in Charts 2, 3 and 4 in Chapter 6. The number was the square root of one percent of the population of the world at that time in 1983.

In the first Iraq war that started in August of 1990, when Saddam Hussein invaded and annexed Kuwait, a coalition of thirty four countries joined in a war to fight Saddam Hussein and liberate Kuwait. Maharishi again called for groups of TM and Yogic flying participants to come to Maastricht in Germany. There were 4,000 TM and Yogic flying participant who came. I told a friend I was going there too, to stop the Iraq war, like Maharishi predicted it would stop. He said, "You must be crazy, you cannot stop the war!"

After just a few weeks, when the threshold of the group of 4,000 was reached, the first Iraq war, then called the Gulf war ended in February 28, 1991 with Kuwait liberated from the forces of Saddam Hussein. The date was just weeks from Maharishi's birthday in January 12, 1991 during which he called to commence the group assembly. It was like when the invincible Unified Field shield was spread over Europe and Middle East, the first Iraq war came to an end barely two months.

"It was just coincidence," my friend exclaimed. It was not a coincidence, the second Iraq war in 2003 lasted more than ten years and now another war going on in Iraq. It cannot be coincidental as other TM assemblies displayed the same results. In a scientific investigation, if the results are the same each time the event are repeated, it becomes a fact. The Unified Field applied in groups together was repeated more than fifty times, the results were the same each time it was repeated - all conflicts end. The discovery of the Unified Field is the greatest fact modern science has found to date. It can balance climates. Peace and prosperity prevail.

I found out the application of the Unified Field is a fact. In the fourth school I had built in a remote, rebel ridden, poor town in the Philippines in 1994, rain fell as soon as 500 students started the TM program. It happened on a warm day in July that season. The rebels vanished in town. Significant signs of economic recovery started to happen. It was not the only one event that I had formed. I witnessed another event which displayed the true fact of the Unified Field in place. In the second TM school I had built in a downtown city of the Philippines in 1988, street killings were common and even the chief of police was gunned downed and shot dead, as soon as the 400 students started the TM program, the killings subsided. The school property increased a hundred-fold as soon as the thirty storey tower penthouse floor level was poured in place. The tower was built on the land where the school was built.

Maharishi was correct when he said that the group practice of TM and Yogic flying can simmer down conflicts, like in 1984, he sent 1,200 TM participants, equal to the square root of one percent of the population of the Philippines to quell down the people uprising at that time in the Philippines. The great rebellion stopped. A new government was peacefully installed. Not a single individual was killed. The earthquake and the typhoons that prevailed at that time, faded away. The volcano that was erupting simmered down. Then President Marcos fell out of power but was safely flown to Hawaii. A new government peacefully took over. In contrast to the rebellion in Libya, Maummar Kaddafi, his grandiose palace destroyed and pulverized to the ground and his treasures looted and stolen. Kaddafi was dragged and killed in the streets by mobs that rose in Libya. What a tragic death and demise of power!

Maharishi called for an assembly in Washington D.C. again in 1994. There were 4,000 participants who came. I was one of them. In just three weeks, crime rates went down to 23%. Washington D.C was then the crime capital of the USA. If a permanent group is formed, no crimes will occur. Maharishi said that heaven on earth will be established.

Today in Syria, where nearly 300,000 people killed and about twelve million people have fled the country. Syria appears completely destroyed by war with half of the population gone! The whole world watch in awe with no easy solution after nearly five years of civil war. Russia came into Syria, with the intent to help the leader of Syria, Assad, to stay in power. Russia has fired thousands of missiles from the air killing hundreds of civilians. Killing is against natural law and takes retribution to whoever violates its law. The cost of oil has gone down reducing Russia's economic status. Who knows, it might be due to all the killings Russia has done including many deaths in Ukraine. As the saying goes, "As you sow, so shall you reap." In the Christian bible, it says, "If someone throws you stone, throw back bread," it works better.

In Iraq, terror groups have risen and have control major towns. Meanwhile, Christians are being beheaded and burned! In Afghanistan, a hospital was bombed killing ten patients and thirty three staff with no concern about the terror groups but to heal the sick. There is a solution to end the conflicts without the use of bombs, missiles or assault guns. I wonder why it is not being used. Maybe because the science of the finest particles of nature described in quantum physics is a recent find. The Unified Field was a recent discovery that still needs to become popular. I know of only two governments that used it, in Cambodia whose government offered Maharishi a land to build a TM school and in Mozambique, Africa where large groups of military practiced TM and Yogic flying in groups together in 1994. What happened in Mozambique is phenomenal. The long two decades of civil war ended. Rain fell in drought areas. Its economy rose 19%. The same rain fell in drought stricken Brazil during the Global Taste of Utopia event in Fairfield, Iowa, USA. When the same assemblies are repeated, the same results prevail.

To me, Maharishi has surpassed Albert Einstein for Maharishis discovery of the Unified Field which Einstein throughout his life pursued but failed to locate the Unified Field. Truly, Maharishi is the greatest scientist to date. Maharishi located the Unified Field within the silent state of the human mind. Maharishi discovered the capability of man to remove problems and suffering from within the silent level of the human mind in large groups together. Man now can create heaven on earth.

Erlinda Case, Author